MIT: Shaping the Future

Ed,
With gratitude for your contribution to this volume + to MIT.

Chuck Vest

MIT: Shaping the Future

edited by Kenneth R. Manning

The MIT Press
Cambridge, Massachusetts
London, England

© 1991 Massachusetts Institute of Technology

All rights reserved. No part of this book may be reproduced in any form by any electronic or mechanical means (including photocopying, recording, or information storage and retrieval) without permission in writing from the publisher.

This book was set in New Baskerville by The MIT Press and printed and bound in the United States of America.

ISBN 0-262-63141-5 (pbk.)

Library of Congress Catalog Card Number 91-66244

Contents

Preface   vii

Contributors   ix

Does Technology Shape the Future?   1
Rosalind Williams

Equality of the Sciences at MIT   9
Robert K. Weatherall

Technology and the Liberal Arts   22
Ann F. Friedlaender

Mary, Theresa, and Elizabeth   32
Cynthia Griffin Wolff

Addictions and Recovery at MIT   46
Eve Odiorne Sullivan

A Personal View of Education   53
Alan V. Oppenheim

Leadership through Science and Technology   62
Michael L. Dertouzos

The University as Quality   74
Alfred R. Doig, Jr.

Toward International Education at MIT   79
Catherine V. Chvany

The Research Library in the Information Age   92
Jay K. Lucker

Electronic Organs of the Mind: Language and Computation for the 21st Century   103
Robert C. Berwick

System Dynamics: Adding Structure and Relevance to Precollege Education   118
Jay W. Forrester

Clean Utilization of Fossil Fuels   132
János M. Beér

An Environment for Entrepreneurs   139
Edward B. Roberts

MIT and Industry: The Legacy and the Future   163
Thomas R. Moebus

The Bear in the Woods   179
Lester C. Thurow

Appendix
Inaugural Address, May 10, 1991   189
Charles M. Vest

Preface

This commemorative volume began as an idea of the committee charged with planning the inauguration of Charles M. Vest as president of MIT. The idea was to present to the new president a collection of views and viewpoints tied to the theme "MIT: Shaping the Future." Such a collection, we hoped, would provide a deeper understanding of the community he has chosen to join and lead.

The question was whether the MIT community, diverse as it is and with the wealth of experience underlying it, would rally to reflect, shape those reflections, and actually put them down on paper. Not knowing quite what to expect, the publications subcommittee issued an appeal for contributions. We are pleased that a resoundingly positive response ensued from many parts of the MIT community.

The essays present a wide range of perspectives on the nature of our mission in education and research, though they cannot and do not attempt to represent the full spectrum of concerns relevant to the MIT experience. There are, for example, no essays on the changing demographics of the MIT student body, the status of engineering education, or the role of ethics in scientific research—all issues of current and ongoing concern. Our goal was not to be comprehensive but to present a sampling derived from voluntary responses within the community. Instead of striving for some common denominator, we chose to present the numerators—a series of individual perspectives and experiences—to help us gain insight into where we are as a community, and where we are going.

The authors speak in their own voices—not as representatives of any department, discipline, or area. They draw on a rich fund of experience as teachers, scholars, critics, and members of an academic community. Their aim is to exhibit the fruits of their work, share ideas, and articulate concerns—and in so doing to acquaint our new president and others at MIT and beyond with their individual perspectives on what our role as a community is and might be. By implication, these essays consider the daunting challenge of taking MIT into the future.

The organizing principle of the volume is loose but not random, intended to underscore the richness and diversity of the authors' approaches. The essays move from those by Williams and Weatherall, which discuss general issues, to the more personal narratives of Oppenheim and Wolff, to the specific research reviews of Berwick and Forrester, back again to a treatment of more general issues identified by Roberts and Thurow. A reading of the volume in reverse order would be just as appropriate. Each essay provides fresh insight on some important issue facing us as an institution and a community.

The members of the publications subcommittee are grateful for helpful counsel, encouragement, and support provided by the Inauguration Committee. Larry Cohen of the MIT Press and Nancy Ferrari provided invaluable editorial service. We extend our thanks to them as well.

The Inauguration Committee

E. Jane Betts, Claude R. Canizares (Chair), Kenneth D. Campbell, Daniel J. Dunn '94, David S. Ferriero, Gayle M. Fitzgerald, Ellen T. Harris, William J. Hecht, Stephen D. Immerman, Henry D. Jacoby, Martha R. Jennings, Norman B. Leventhal, Kathryn W. Lombardi, Kenneth R. Manning, Christian J. Matthew, Laura B. Mersky, Laura C. Moore '91, Conor Moran, Mary L. Morrissey, Leo Osgood, Denise A. Purdie '91, Martin F. Schlecht, Constantine B. Simonides, Kaiteh Tao '94, Lyna L. Wiggins

The Publications Subcommittee

Kenneth D. Campbell, Naomi F. Chase, Kenneth R. Manning (Chair), Kathryn A. Willmore

Contributors

János M. Beér
Professor of Chemical and Fuel Engineering

Robert C. Berwick
Associate Professor of Computer Science and Engineering and Computational Linguistics

Catherine V. Chvany
Professor of Russian

Michael L. Dertouzos
Professor of Computer Science and Electrical Engineering
Director of Laboratory for Computer Science

Alfred R. Doig, Jr.
Assistant Dean for Resource Development, School of Engineering

Jay W. Forrester
Germeshausen Professor of Management, Emeritus

Ann F. Friedlaender
Class of 1941 Professor of Civil Engineering and Economics

Jay K. Lucker
Director of Libraries

Thomas R. Moebus
Director of Industrial Liaison Program

Alan V. Oppenheim
Distinguished Professor of Electrical Engineering

Edward B. Roberts
David Sarnoff Professor of Management of Technology

Eve Odiorne Sullivan
Senior Editorial Assistant, Laboratory for Nuclear Science

Lester C. Thurow
Dean of Sloan School of Management
Professor of Management and Economics

Robert K. Weatherall
Director of Office of Career Services

Rosalind Williams
Class of 1922 Associate Professor of Writing

Cynthia Griffin Wolff
Class of 1922 Professor of Literature

MIT: Shaping the Future

Does Technology Shape the Future?
Rosalind Williams

In his famous 1959 lecture titled "The Two Cultures," C. P. Snow struggles with the problem of defining what he calls "the scientific culture." Despite obvious differences in politics, religion, class, and professional specialization, Snow contends, scientists share "common attitudes . . . common approaches and assumptions. . . . If I were to risk a piece of shorthand, I should say that naturally they had the future in their bones." As a corollary, Snow asserts that the other culture—which includes nonscientific intellectuals, especially literary ones—is oriented to tradition and the past.[1]

I would argue that Snow has it backwards; that engineering and science are highly traditional enterprises, while humanists serve as "the antennae of the race," to borrow a nice phrase from Ezra Pound. The histories of science and technology overflow with examples of the hold of the past. As MIT's Thomas Kuhn has shown so persuasively, most scientists do their work within the limits of inherited paradigms. Historians of technology, for their part, have repeatedly demonstrated the cultural power of received models (the automobile perceived as a "horseless carriage," early iron columns fluted to resemble stone ones) and the fundamental conservatism of most practitioners. The history of military technology in particular provides striking examples of humanists who were far more alert to the future than engineers. At a time when British commanders were only beginning to realize that tanks had superseded horses in combat, H. G. Wells foresaw the use of atomic weapons. In a similar way, contemporary writers of cyberpunk science fiction, not engineers, have most clearly recognized the problem of aging, decaying infrastructures in an advanced industrial society.

In short, it is not difficult to respond to Snow's anecdotal evidence on the same level, if we want to pit "the two cultures" in a competition to see which is the most hidebound. The issue Snow raises, however, is far too important to remain on that casual level. The assumption that the scientific culture has the future in its bones is closely related to larger, widely held, but little examined assumptions about historical causality. History develops as it does (so goes the conventional reasoning) because scientific rationality leads to technological innovations, which in turn lead to social progress: time plus science and technology equals the future. As a last link in this line of reasoning, we might assume that since MIT is in the forefront of scientific research and technological applications, MIT powerfully shapes the future.

The issue is not whether MIT is in the scientific and technological vanguard: That much is indisputable. The issue is whether technological change, bred of scientific discoveries, is indeed the dominant force driving historical change. The belief that it is—a belief in technological determinism—cuts across the usual political boundaries. Karl Marx is often described as a technological determinist (in a famous passage he declared, "The hand-mill gives you society with the feudal lord; the steam-mill, society with the industrial capitalist").[2] But capitalists too can be fervent technological determinists. For example, in a June 1989 speech Ronald Reagan predicted that "the Goliath of totalitarian control will rapidly be brought down by the David of the microchip."[3] A philosophy of history that is so widely held, and that has so many implications for public policy, deserves to be examined with some rigor and care.

While there are many varieties of technological determinism, they can all be condensed into a three-word logical proposition: "technology determines history." The best-known scholarly overviews of the subject—those by Donald MacKenzie, Langdon Winner, Robert Heilbroner, and Political Science/STS graduate student Bruce Bimber, among others[4]—sooner or later get around to analyzing each of these three key terms.

What is technology, for starters? Is it synonymous with "machines," or should we define it more broadly as "means of production"—as Karl Marx did—and if so, what does that mean? In many of these discussions, it should be noted, there is an implicit assumption that certain technologies are more primary, more significant, more determinative if you will, than others—steam engines and automobiles, for example, rather than domestic and military technologies. This division of production into primary, secondary, and perhaps even tertiary sectors can be the source of considerable confusion when applied to a late capitalist economy, where the proportion of primary production in the traditional sense of sustaining life is so small. As the late cultural critic Raymond Williams observed, "By the time you have got to the point when an EMI factory producing discs is industrial production, whereas somebody elsewhere writing music or making an instrument is at most on the outskirts of production, the whole question of the classification of activities has become very difficult."[5]

As for the crucial verb "determines," it can be constructed in either a "hard" or a "soft" way. The meaning is further complicated because in science itself, from which concepts of determinism are so largely borrowed, modern investigators prefer to speak of probable trends (perhaps highly probable, but still probable) rather than inevitable results. If we substitute the verb "shapes," the problem remains. Does "shaping" refer to a general diffuse influence, or a much stronger activity?

Finally, there is "history," the most problematic term of all, since it is not at all self-evident how to characterize a historical outcome. Consider, for example, the popular late 19th-century predictions that the advent of electrical power would lead to a decentralized electronic utopia.[6] Has history proved these predictions (technological cause, historical result) right or wrong? Wrong: we have urban congestion, densely packed skyscrapers above, traffic jams below—or right, if you look at suburbs, telecommuting, and the decentralization of consumption through domestic appliances. You can also say that technology produces historical contradictions, which may be a more accurate response, but one that can be evasive if you use "contradictions" simply to mean a "mixed outcome."

In defining "history" in relation to technological determinism, we are in danger of going around in logical circles. We tend to do implicitly what Robert Heilbroner does explicitly at the beginning of his classic article "Do Machines Makes History?"—that is, to define history in terms of a socioeconomic order, rather than in terms of, say, political or diplomatic events. This way of defining history is in itself a result of priorities that are technology-based (if not technologically determined). Technology decisively entered the study of history in the Enlightenment and early 19th century. In response to the great technological event of that epoch—the overwhelming and unprecedented increase in productivity—the concept of technological progress was gradually extrapolated to history as a whole, and history became redefined as the record of socioeconomic progress. In other words, for those of us living in the modern age, history is almost by definition a technology-driven process.

Even this brief analysis of the three terms "technology," "determines," and "history" suggests both the value and the limitation of textual deconstruction. Plainly we need to unpack these huge terms, to analyze language that is all too often repeated uncritically. On the other hand, we may also end up juggling categories, tossing around abstractions that eventually float free from connections to the social world of material interests and intentions. To paraphrase Heilbroner, we have to think of historical determinism itself in historical terms, and even in political terms. What interests are involved here? Why would any individual or group be motivated to assert technological determinism—hard, soft, or otherwise? More specifically, I would suggest asking this crude but provocative question: "Would any woman have come up with this idea?" To be somewhat less crude, is a theory of technological determinism at all compatible with a feminist understanding of history?

In many cases, interests are best revealed not by what people assert, but by what they implicitly deny. To affirm that technology drives history is to say that God does not. Marx in particular was denying that history is directed by God, or by some Hegelian or other spiritual ideal. But technological determinists do more than slay God the Father; they also slay Mother Nature, or at least declare her death. Marx is by no means alone in declaring that nature has ceased to be an independent force in history, that its role has been displaced by the "second nature" of humanmade artifice. A whole epoch of bourgeois thought is involved here, an epoch of "technocratic consciousness" (Jürgen Habermas's term) which affirms that history is made by humanity as nature is made by God. History is declared to be part of what Sir Francis Bacon called "the human empire." Thus technological determinism declares humanity's liberation from spiritual and natural necessity alike.

To declare independence is to make a political statement—a revolutionary statement—and technological determinism can be, and has been, appealed to as a revolutionary force. Once again, Karl Marx is the obvious example, but one that must be appealed to with caution; if Marx is a technological determinist at all, he is an exceedingly subtle and ambivalent one.[7] Still, Marx did announce his determination to change the world, and, paradoxically enough, he proposed that what appears to be determinism actually liberates the proletariat to assume its historical destiny. If capitalist politics or thought had any significant role in shaping history, then maybe through some manipulations or concessions or new ideas capitalism could save itself. But when its fatal contradictions arise from its mode of production (and however one wants to interpret that phrase, it is clearly based in technology), then collapse cannot be averted. Revolution is inevitable precisely because technology is largely "out of control."[8]

This assertion of technological determinism as a form of revolution is the link, strangely enough—very strangely enough— between the philosophy of Karl Marx and that of Ronald Reagan. To call Reagan a thinker is generous, but just because his mind is relatively unsubtle he faithfully echoes themes that reverberate in the social circles he frequents. In these circles too, evidently, technology generates revolution. The microchip will bring down totalitarian control by undermining centralized authority; the communications revolution will also be a political revolution. If for Marx technological determinism makes revolution inevitable, for Reagan it makes revolution unnecessary.

Obviously this connection between technological determinism and political revolution does not hold for all cases, but it does suggest that we should pay more heed to the intentions and interests of the proponents of the idea—not only those who talk about the technological determinism, but also those who can act upon it. It has often been noted that the technologies least responsive to human will—the ones with the most momentum, to use the term favored by historian of technology Thomas P. Hughes—are ones of large scale, size, and complexity. In that case, we should look at people who choose to invest in large, complex technologies, and consider that they may do so quite deliberately in order to create technological determinism. As Hughes also reminds us repeatedly, engineers build values into their designs. They can design systems that are highly responsive to human control—or ones that are not. Fate can be engineered. To take an obvious example (and here I am quoting essayist Wendell Berry), "We may choose nuclear weaponry as a form of defense, but that is the last of our 'free choices' with regard to nuclear weaponry. By that choice we largely abandon ourselves to terms and results dictated by the nature of nuclear weapons."[9]

In "Do Machines Make History?" Robert Heilbroner has suggested that the concept of technological determinism works best in this historical epoch where "forces of technical change have been unleashed, but when the agencies for the control or guidance of technology are still rudimentary." I am just adding a reminder that some groups profit mightily by the absence of social controls. In a way, I'm paraphrasing the rightwing slogan, "Guns don't kill people, people do," to argue that in many cases "Machines don't make technological determinism, people do."

All this brings us back to the apparent dissonance between technological determinism and a feminist understanding of history. Few women would be in the position to create technological determinism in the way just described, and few women would be so ready to proclaim the liberation of human history from nonhuman nature. Their history, after all, has been largely involved with the socialization of the organic. I realize I am on a slippery slope here, since so much feminist energy has correctly been spent in opposing natural determinism, in protesting that biology is not destiny. In the striking phrase of Ynestra King, "Women have been culture's sacrifice to nature,"[10] and I have no wish to drag women back to that sacrificial altar. I wish rather to assert that there is an inextricable connection between humankind and nonhuman nature; the problem lies not with asserting this connection, but with restricting it to the category of a "feminist" issue.

There is certainly room in the Marxist tradition for a historical materialism that comprises natural forces as well as technical ones. I still remember sitting in a freshman class in medieval history at Wellesley College, listening raptly as the professor expounded Karl Wittfogel's theory of hydraulic societies—Wittfogel being an erstwhile Marxist who explained the authoritarian character of Oriental societies by their need for complicated water management systems. What made the theory so appealing to a freshman is also its weakness: its simplistic and universalizing environmental determinism. But in a world of rapidly depleting aquifers, ozone holes, and global warming, Wittfogel's theory looks less simplistic than assertions of humanity's triumph over natural necessity.

Still, assertions of humanity's triumph are not all wrong either. Human imperialism is a fact: Our population and technologies have reached such a scale that they have intertwined with natural systems. Nature may still be a force, but it is no longer an independent one. We human beings have not escaped from nature, but neither has it escaped from us. Is global warming a technical event or a natural one? We can no longer discern the difference. How about El Niño? Ozone depletion? When we human beings can kill entire lakes and forests through industrial fumes and burning, when we can alter food chains with oil spills and scatter radioactivity across central Europe, then we dwell in an environment where natural and technological processes have merged. This new hybrid environment may not determine the fate of humankind, but it will profoundly and decisively affect our future.

How, then, can MIT shape the future in this changed environment? Certainly the problems it presents require technological solutions, and more particularly the creation of technologies that are, by their scale and complexity, inherently flexible rather than determinative. Just as certainly, however, the solutions are not only technological. Profound changes in human behavior are also required, changes that emerge from transformed values and that lead to transformed institutions; changes that involve redefining the good life and rethinking the meaning of social justice. All the technological inventions in the world will not help our situation, or will only make it worse, if they are not complemented by social inventions.

The degree to which MIT shapes the future depends on what we emphasize in our name. If we stress that this is the Massachusetts Institute of *Technology*, we may end up with a subordinate role in bringing about truly significant historical changes. But if we emphasize that we are part of the Massachusetts *Institute* of Technology—a social invention with a proud heritage, a place where people work together to create a society of diversity, equity, and justice—then MIT will indeed play a leading role in shaping a better future.

Notes

1. C. P. Snow, *The Two Cultures and the Scientific Revolution*, The Rede Lecture 1959 (New York: Cambridge University Press, 1959), 11.

2. Karl Marx, *The Poverty of Philosophy* (New York: International Publications, 1971), 109.

3. Quoted by Kenneth Keniston in *STS News* (October/November 1989), 1. Reagan was addressing the English Speaking Union in London.

4. Donald MacKenzie, "Marx and the Machine," *Technology and Culture* 25 (July 1984), 473–502; Robert Heilbroner, "Do Machines Make History?" *Technology and Culture* 8 (July 1967), 333–345; Langdon Winner, *Autonomous Technology: Technics-out-of-Control as a Theme in Political Thought* (Cambridge: MIT Press, 1977); and Bruce Bimber, "Karl Marx and the Three Faces of Technological Determinism," STS Working Paper No. 11, MIT, and *Social Studies of Science* (May 1990).

5. Raymond Williams, *Interviews with the New Left Review* (London: Verso, 1981), 353.

6. These predictions are ably summarized by James W. Carey and John J. Quirk, "The Mythos of the Electronic Revolution," *American Scholar* 39:2 (Spring 1970), 219–240; 39:3 (Summer 1970), 395–424.

7. See especially the articles by MacKenzie and Bimber cited above.

8. See the book by Langdon Winner (a former member of the STS Program at MIT) cited above.

9. Wendell Berry, "Property, Patriotism, and National Defense," in *The Contemporary Essay*, 2nd ed., ed. Donald Hall (New York: St. Martin's Press, 1989), 56. This text (and the particular essay) are used in the MIT class 21.735, "Writing and Reading the Essay."

10. Ynestra King, "Healing the Wounds: Feminism, Ecology, and Nature/Culture Dualism," in *Gender/Body/Knowledge*, ed. Alison M. Jaggar and Susan Bordo (Brunswick, NJ: Rutgers University Press, 1989), 129.

Equality of the Sciences at MIT
Robert K. Weatherall

Famous for its teaching and research in economics, management, political science, planning, and linguistics as well as for its programs in engineering and the natural sciences, MIT has a hard time describing itself. Although obviously more than an engineering school, it does not see itself as a university—if by a university is meant such an institution as Oxford, Munich, or Michigan. "University" points to something more comprehensive than what MIT wants to be. For a while MIT played with the phrase "a university polarized around science." But this begs too many questions. First of all, what science, or, rather, which sciences? Without more specification the door is open to all the *wissenschaften* of von Humboldt's ideal university—to every discipline which aims at scientific objectivity. And what is meant by "polarized"? The possible analogies from physics suggest a refracting or bending process of some sort. Is that the character of an MIT education?

In the absence of an accepted description of what MIT is about, the Institute community falls back on a set of venerable assumptions. According to these, MIT is first and foremost what its name implies—an institute of technology—and its primary mission is to prepare students for careers in the natural sciences and engineering. These subjects are central to the Institute. Most central of all are mathematics and physics. Other subjects are more or less fundamental, "hard" or "soft," more or less scientific, more or less worth pursuing, depending on how far they stand from mathematics and physics.

Join any lunchtime conversation in which the Institute is the topic and you are likely to find these ideas surfacing in one form or another. They crop up in discussions of the undergraduate curriculum, in the advice faculty members give their students, and students give each other, on what to study, in the speeches of senior administrators, and in the reports on its activities the Institute gives the public.

A vivid recent example was this year's debate in the faculty on making biology a core requirement for undergraduates alongside mathematics, physics, and chemistry. All the arguments pro and con took certain things for granted. No one questioned, for example, that MIT was a "technological university," a phrase used frequently in the discussion. No one doubted the appropriateness of a "science core." (Never mind that a core requirement of any sort is increasingly rare at American colleges.) No one disagreed that if another subject should be added to mathematics, physics, and chemistry, biology was next in line. Everyone agreed that the revolutionary advances in the subject in the last 40 years had brought biology to "maturity." The only serious question at issue was whether biology should be added to the core, compelling students to take a specific course in biology, or made a mandatory distribution subject, in which case students would have a choice of which biology course to take.

The arguments for biology would have had to be put differently if they had had to contend with a different image of MIT. Other subjects represented at the Institute have been transformed since the core curriculum was put in place and could claim, like biology, that they have become mature sciences. Economics, political science, and management are good examples. If a core requirement makes sense at MIT, why should it include only natural science subjects? If biology, why not economics? If social scientists need to be well-grounded in the natural sciences, why do natural scientists and engineers not need an equal grounding in the social sciences? Does not the rationale for a "science core" need to be reexamined, perhaps, now that the Institute graduates social scientists as well as natural scientists and engineers? Characteristically, questions such as these were not broached in the debate, and the discussion stayed in the usual well-worn channels.

The Institute's ranking of the sciences from more to less scientific recalls the memory of Auguste Comte, who distinguished between "positive" science and theological and metaphysical system building, ranked the positive sciences in a hierarchy according to their generality, with astronomy and physics the most general, and argued that a student needed to study them in order. "Physical philosophers cannot understand Physics," he wrote in his *Cours de Philosophie Positive* (1830–1842), "without at least a general knowledge of Astronomy [i.e. of Newton's laws]; nor Chemists, without Physics and Astronomy; nor Physiologists, without Chemistry, Physics, and Astronomy; nor, above all, the students of Social philosophy, without a knowledge of all the anterior sciences."[1]

He gave the *Cours* the tone of a manifesto: "Let us have a new class of students, suitably prepared, whose business it shall be to take the respective sciences as they are, determine the spirit of each, ascertain their relations and mutual connection, and reduce their respective principles to the smallest number of general principles, in conformity with the fundamental rules of the Positive Method. At the same time, let other students be prepared for their specific [employment] by an education which recognizes the whole scope of positive science."[2]

Comte's spirit has hovered over the Institute since the beginning. In his 1846 "Plan for a Polytechnic School in Boston," William Barton Rogers proposed a school with two departments, a "scientific department" teaching physics and chemistry to lay a "groundwork . . . in general physical laws," and a "practical department" offering instruction in specific fields of technical work. "I doubt not," he wrote, "that such a nucleus-school," supported by the "knowledge-seeking" community of Boston, would "finally expand into a great institution comprehending the whole of physical science and the arts, . . . and would soon overtop the universities in the land in the accuracy and the extent of its teachings in all branches of positive knowledge."[3]

His plan does not make Rogers a Comtist, but the idea of the two departments and the reference to "positive knowledge" strike a chord. Few of Rogers's scientific contemporaries were in the habit of using the positivist vocabulary. While Comte's positive philosophy had a significant impact on 19th-century opinion, his notion that Christianity should give way to a religion of humanity (with its own priesthood), and his autocratic ideas for the organization of industrial society (spelled out in minute detail), led most thoughtful souls to view him as an inspired crank. Few contemporaries, however much they shared his faith in science and gave credit to his analysis of scientific methodology, were prepared to call themselves positivists. John Stuart Mill, who praised his *Cours de Philosophie Positive* when it first came out as "far the greatest [work] yet produced on the Philosophy of the Sciences," commented later that "the mode of thought expressed by the terms Positive and Positivism [was] better known through [its] enemies than through its friends."[4]

Asa Gray, who brought the same factual cast of mind to botany as Rogers brought to geology and had no hesitation, just as Rogers had none, in accepting Darwin's *Origin of Species* as a landmark contribution to science, considered positivists akin to materialists.[5] T. H. Huxley, whom Rogers knew and admired, wrote flatly in 1869: "Appeal to mathematicians, astronomers, physicists, chemists, biologists, about the 'Philosophie Positive,' and they all, with one consent, begin to make protestation that, whatever M. Comte's other merits, he has shed no light upon the philosophy of their particular studies."[6]

Rogers clearly did not mind sounding like a positivist. In a physics textbook he wrote and had printed in 1852 for his classes at the University of Virginia he devoted the first 20 pages to mapping "the entire body of positive science" as he saw it. It comprised, he wrote, "two great classes of truths": those "derived from consciousness, self-analysis, and the study of intellectual phenomena in general," which constitute "mental science," and those "derived from the observation of material bodies, and the phenomena of the external world in the course of physical events," which constitute "physical science."

Rogers did not need to call the latter "physical science." "Natural science" would have done just as well. He chose "physical" advisedly. In his view, all the forces operating in the natural world, "mechanical, chemical, and vital," were "physical" forces. The vital forces manifested in living things, he declared, "are the mechanical and chemical forces proper to matter in a condition of organization." He hoped that "by higher induction" it would eventually be possible "to unite them all under one common principle of force."[7]

In 1855, in an address to the Williams College Lyceum of Natural History, he strove to assure his audience that "the growth of positive science" was not unfriendly "to poetical and spiritual conceptions of the material world." Is there anyone, he asked, who can see "in the material laws which control and harmonize this universe, aught lower or less spiritual than the thought of Infinite Wisdom and the handiwork of Infinite Power?" But after this rhetorical flourish he took care to tell the Williams students what he had wanted his students at Virginia to understand, that "in the state of development which they have now reached, each of the great departments of Natural History is brought into close connection with the purely physical sciences, each has borrowed from them valuable methods and instruments of research, and each invokes the aid of physical laws and forces as part of the machinery by which the phases and activities of organic things are to be explained.... Let us take care that we do not lose ourselves in that mysticism which imparts intelligence and prescience to embryonic cells and organs."[8]

Four years later the *Origin of Species* was to shake the whole fabric of natural history. The book challenged Comtist as much as vitalist ideas. The new view it presented of living things owed nothing to physics and chemistry or to vital forces. The book might be persuasive in arguing that species were not variable, not immutable, that existing species had descended from common ancestors, and that competition and natural selection had determined which survived, but if true these were shakier truths than the demonstrated truths of physics and chemistry. Darwin himself knew that he would not persuade everyone: "any one whose disposition leads him to attach more weight to unexplained difficulties than to the explanation of a certain number of facts will certainly reject the theory." He looked to the future, "to young and rising naturalists, who will be able to view both sides of the question with impartiality."[9]

Although the theory did not fit his map of the positive sciences, Rogers had no hesitation in taking Darwin's side in the great debate that flared up on both sides of the Atlantic, defending Darwin vigorously on the platform and in print. But he stated his position carefully: "The great question, or rather questions, treated of by Mr. Darwin, although by no means new to science, have never before been presented in a form and connection bringing them so clearly within the pale of inductive reasoning. The discussions to which they will give rise, if pursued in the spirit in which Darwin has conducted his inquiries, cannot fail to stimulate observation and to extend the limits of positive knowledge."[10]

How Rogers came by his positivism is an intriguing question. He never discusses his positivist ideas or the influences which shaped them. If he ever read Comte he does not say so. All we know is that the four Rogers brothers, William, Robert, James, and Henry, all of them scientists, were eager readers of the English periodical press and referred each other to interesting articles. Comte's *Cours de Philosophie Positive* was first brought to the attention of the English public by a long and enthusiastic article in the *Edinburgh Review* in 1838.[11] Whether the brothers noticed the article or not, William and Henry became warm admirers of the author, the distinguished Scottish physicist and encyclopedist David Brewster. "I could almost have knelt down to ask his scientific benediction," William told Henry after meeting him in 1849.[12]

Rogers undoubtedly read John Stuart Mill's laudatory references to Comte in his *System of Logic.* He obtained a copy of the second edition in 1847. But Rogers was already talking like a positivist before then, and Mill, while praising Comte, did not adopt his vocabulary.[13]

It is also possible that we have been barking up the wrong tree, that Rogers's positivism did not come from reading this or that, but came with his dream of founding an American equivalent of the French Ecole Polytechnique. Unfortunately he tells us as little about the shaping of his educational ideas as he does about the development of his philosophy of science. His 1846 plan for a polytechnic school in Boston emerges from nowhere, with no preliminaries. No doubt the Ecole Polytechnique was his model, but he does not refer to it in the plan, or anywhere else in his surviving papers. Unlike Alexander Dallas Bache, who visited the school along with others in Europe before taking on the presidency of Girard College in Philadelphia (and wrote a report on his visit), Rogers does not appear to have included it in his visits to Europe.[14]

But we have to believe that Rogers was familiar with the school, and was greatly impressed. Established by the French government in 1794, it was the most famous of several government-run schools of engineering in France, which together made France the world center of engineering education in the first decades of the 19th century. Laplace, Arago, Lagrange, Cauchy, Fourier, Poisson, Gay-Lussac, and Ampere were among those who taught there.[15] It was a school where the mathematical-physical sciences were enthroned in glory. Comte attended the school and made his living as an examiner there. It is not difficult to see his positivist philosophy as a gathering-up and working-out of ideas implicit in the school, of *l'esprit polytechnicien.* If Rogers was a positivist, it may be simply for the same reason Comte was one: They were inspired by the same institution.[16]

It is a pity that Rogers is not better remembered at the Institute he founded, for much about him is in keeping with the place. His father was professor of natural philosophy and chemistry at William and Mary, and his early education was largely under his tutelage. When he was 20, his father had him translate a French textbook on calculus for the use of his younger brother Henry.[17] A workaholic, he poured out a stream of papers in not one but several sciences. He was fond of poetry and history but was careful not to let them interfere with his scientific pursuits. When he tried his hand at evocative writing he tried too hard, as one sees in his address at Williams. Of the famous New England writers who became his fellow-citizens when he moved from Virginia to Massachusetts, only Longfellow is mentioned in the two-volume *Life and Letters* his wife compiled at his death. He reminded her of the traveller in Matthew Arnold's poem, "The Grande Chartreuse," who finds that his 19th-century upbringing has set him apart from the pious monks of the abbey:

For rigorous teachers seized my youth,
And purged its faith, and trimmed its fire,
Showed me the high white star of Truth,
There bade me gaze, and there aspire.[18]

Rogers, like Comte, was inspired by the forward march of the physical sciences in the face of popular ignorance, political upheavals, and theological and metaphysical controversy. Both men were deeply impressed by the sweeping power of the laws of physics and chemistry. They were truths that could stand up to any theological or metaphysical assertion. The proper goal of other disciplines was to arrive at truths of equal power.

Both men also recognized that the industrial revolution going on around them was the dawn of a new age. They took the side of industry against the old order, and saw the advance of positive science and the growth of industry going hand in hand.

Comte, who was horrified by the continuing turmoil in European society in the aftermath of the French revolution, believed that positive science, properly pursued, could create a rational social order. Indeed, his principal aim in writing the *Cours de Philosophie Positive* was to pave the way for a "social physics."[19]

Rogers was more modest in what he expected from science, but he, too, was interested in the problems of society and thought they could yield to scientific methods of enquiry. In 1865 he played a leading part in establishing an American Association for the Promotion of Social Science, getting elected as its first president. Its aims were "to aid the development of Social Science, and to guide the public mind to the best practical means of promoting the Amendment of Laws, the Advancement of Education, the Prevention and Repression of Crime, the Reformation of Criminals, and the Progress of Public Morality, the adoption of Sanitary Regulations, and the diffusion of sound principles on questions of Economy, Trade, and Finance." A specific concern was "Pauperism, and the topics related thereto; including the responsibility of the well-endowed and successful, the wise and educated, the honest and respectable, for the failures of others." The association hoped to bring together "societies and individuals now interested in these objects, for the purpose of obtaining by discussion the real elements of Truth [and] treating wisely the great social problems of the day."[20]

The association took its inspiration from a similar association in Britain whose members included such noted reformers as Florence Nightingale, Lord Shaftesbury, and Edwin Chadwick. Chadwick became a corresponding member of the American association, as did John Stuart Mill, and also Rogers's brother Henry, who had travelled to Britain in 1832 to join the Owenite movement for factory reform and had settled there in 1857, becoming professor of natural history at the University of Glasgow. The American members included the businessman and writer on economics Amasa Walker, whose son Francis was to succeed Rogers as president of MIT.[21]

Would Rogers have put together a different plan for MIT if his interest in social issues had developed earlier? Perhaps so, but it is far from certain. At MIT his central concern continued to be the education of engineers and industrial managers who were solidly grounded in mathematics and the physical sciences. When the corporation found the funds for a new faculty position in 1872, it was to pay for a professor of logic and the philosophy of science. The professor, George H. Howison, had recently published a well-received treatise on analytical geometry and an extended paper on the different parts of mathematics and their mutual relations. Once appointed, he was also asked to teach economics, but that was an afterthought.[22]

Since Rogers's day the industrial revolution has run its course in the western world, and other sectors of the economy have become as complex, and as dependent on a highly skilled work force, as the manufacturing industry. Today, services constitute over 70 percent of the American gross national product and provide employment to 75 percent of the work force.

Services which were delivered person-to-person are now delivered by organizations matching industrial corporations in size. And organizations have developed their own need of services, services as complicated as any. In this world of organizations the individual citizen finds himself involved at every turn in impersonal, bureaucratic relationships. The material conditions of life have greatly improved since the 19th century, but at a cost. The political, economic, and social upheavals of the 19th century have been followed by even greater upheavals in the 20th century.

The task of running and monitoring the organizations which hold today's world together, of financing them, of ensuring they are doing their work efficiently and well, of protecting the interests of the individual men and women with whom they interact—customers, patients, students, pensioners, neighbors, tax-paying citizens, employees—has stimulated the development of an array of social sciences, pure and applied, from anthropology, sociology, and political science to business management, welfare economics, operations research, land-use planning, organizational behavior, and the rest.

The social sciences have worked out their own methodologies, whatever suited them best. What makes them all sciences is their spirit of unfettered enquiry and their respect for data, for scientific objectivity. Some have followed the model of the mathematical sciences, but most have not. The distinguished social science faculty at the University of Chicago wrote in 1954: "We think of [the social sciences] as a field without boundaries, grading into surrounding fields which have different foci—such as appreciation, action, scientific understanding of other matters. We wish to cultivate all of our borders. On one frontier we touch biology; on another we reach toward the models of the physical sciences and mathematics and more generalized concepts of behavior; and on a third we touch the humanities. . . . In other directions we push forward into new knowledge and theory of human behavior as problems of social action are considered. In all directions we reach for tools and techniques wherever they may be provided."[23]

The social sciences work with probabilities, not certainties. Human behavior cannot be predicted like the orbit of a planet. But probability has also taken the stage in the natural sciences. We have already seen it in Darwin's theory of evolution, but quantum theory has also made it a central concept in physics. In such complex fields of natural science research as the warming of the earth's atmosphere the only reasonable approach is probabilistic.

At the same time, the natural sciences are no longer doing battle with the absolutes of theology or metaphysics. Theology and metaphysics have retreated from the field. The chief threats to science today come from a disenchantment with the fruits of science on the part of the public, from a perception that scientists are an interest group like any other and are not impartial gods above the fray, and from public ignorance. All the sciences face these problems.

Rogers was one of many founders of engineering schools in the last century who threw up their hands at the medieval heritage of existing colleges and universities and decided to set up an entirely new sort of institution. Since then—indeed the process had already started in his lifetime—the model of the university has been transformed. Today, almost all the schools of science and technology with which MIT likes to rank itself—and with which others like to compare it—are schools within universities. This is true worldwide, from Cambridge and Munich in Europe, to Michigan, Berkeley, and Stanford in the United States, to the University of Tokyo in Japan. In these universities one science may have more prestige than another—because of the quality of the faculty, or the quality of the students it attracts, or the importance employers attach to it—but it is not considered scientifically more worthy. In the eyes of the university the sciences are equal.

MIT has flourished pursuing its own path and is admired the world over as a teaching and research powerhouse, but, except in engineering education, it is seldom taken as a model by other institutions. We are seen as having a different educational philosophy, and our educational prescriptions are therefore thought irrelevant. While fewer and fewer colleges retain a core curriculum, no one notices our science core. In the ongoing debate on the place of western culture in the humanities requirements of American colleges, our humanities requirements evoke no interest. It will undoubtedly be the same with the faculty's decision to add biology to the science core. In matters of general education, MIT is a voice in the wilderness.

If this were the only harm done by our attachment to Comte's ranking of the sciences we might live with it, but there are other consequences which are more serious. The most serious is that it discourages students from choosing majors at the "soft" end of the spectrum, and it discourages many more from applying to MIT in the first place. The quality of our faculty in the social sciences should attract as many students as are attracted by the Institute's reputation in the natural sciences and engineering. They go elsewhere because they see the core science requirement as a needless hurdle. It is not a foolish perception on their part. One does not need to have taken the core science subjects to study the social sciences at MIT as a graduate student. Applicants can be excused if they see the requirement as one prescribed by the engineers and natural scientists on the faculty, not by the social scientists.

The requirement creates a filter that probably also turns away many would-be engineers and natural scientists. They want to mix with students who have other interests, and they see more opportunity for that at a university. Many probably do not even consider the Institute. If they applied here, we would rate them highly. Many observers have commented that engineering students are more varied in their interests than they used to be. It is difficult to measure how many bright candidates with such interests we are losing to other schools, but perhaps an indication is the drop since the 1970s in the percentage of applicants with top-of-the-line scores (math and science scores above 750, verbal and English scores above 650). In the early 1970s they constituted 7.6 percent of the applicant pool; this year the figure was 5.6 percent.[24]

Comte thought it a virtue that in "astronomy, physics, chemistry, and physiology there is no such thing as liberty of conscience." The other side of the coin is that an education centered on these sciences is not conducive to the philosophical give-and-take that should be the pride of a university. One can have too little theology and metaphysics. The Institute's dining halls and well-kept lawns are the scene of too few conversations worth remembering on the great issues of life, on books, music, politics, art. We are so used to the Institute's style that we take the lack for granted. We are surprised if we run into a student conversation at MIT of the sort Herbert Simon, social scientist and Nobel laureate, recalls from his student days at Chicago in his memoir, *Models of My Life*. The conversations John Maynard Keynes remembers at Cambridge in his essay, "My Early Beliefs," are even further from the MIT experience. It means that MIT students enter the world with a limited feeling for the range of human creativity and expression, less able than they might be to present their own ideas, less qualified to lead.[25]

Edwin Land, who knew MIT well, worried about the "authoritarianism" of the core curriculum in a notable lecture he gave in Kresge: "The role of science is to be systematic, to be accurate, to be orderly; but it is certainly not to imply that the aggregated, successful hypotheses of the past have the kind of truth that goes into a number system." He saw a disposition to treat the undergraduates as "young and immature," when what was needed was to help them "find ways of handling the intricacy of our culture."[26]

Comte is long dead and his positive philosophy is scarcely noticed any more by students of intellectual history. It is time MIT banished his ghost. The Institute should discard its linear ordering of the sciences and consider them, rather, as forming a circle—a circle in which the social sciences rub shoulders with the natural sciences on equal terms. The general requirements for the bachelor's degree should be revised in that spirit. The professional character of an MIT education would remain unchanged. For a vision of this new MIT, think what it would be like if the MIT we know and the equivalent of the London School of Economics were merged into a single institution. What an institution to help America and the world tackle the technical and social problems of the 21st century!

Notes

1. Auguste Comte, *Positive Philosophy*, freely translated and condensed by Harriet Martineau, 1855, republished with an introduction by Abraham S. Blumberg (New York, 1974), 48.

2. Ibid., 32.

3. Emma Rogers, *Life and Letters of William Barton Rogers* (Boston, 1896), Vol. 1, 420, 421.

4. John Stuart Mill, *A System of Logic, Ratiocinative and Inductive* (London, 1843), Book III, Chapter V, #9, quoted by W. M. Simon, *European Positivism in the Nineteenth Century* (Ithaca, 1963), 275; John Stuart Mill, *Auguste Comte and Positivism* (Ann Arbor, 1961), 2.

5. Letter from Asa Gray to J. D. Hooker, quoted by A. Hunter Dupree, *Asa Gray* (Cambridge, 1959), 297.

6. T. H. Huxley, "The Scientific Aspects of Positivism," *The Fortnightly Review*, Vol. V (New Series), No. 30 (June 1869), 658.

7. William Barton Rogers, *Elements of Mechanical Philosophy, for the use of the Junior Students of the University of Virginia* (Boston, 1852), 2, 19.

8. William Barton Rogers, *Address before the Lyceum of Natural History of Williams College, August 14, 1855* (printed in Boston, 1855), 9, 10, 19.

9. Charles Darwin, *On the Origin of Species By Means of Natural Selection* (New York, 1889), 422, 423.

10. Rogers, *Life and Letters*, Vol. 2, 27.

11. *Edinburgh Review*, Vol. 67, No. 136 (July 1838).

12. Rogers, *Life and Letters*, Vol. 1, 308.

13. Rogers's copy of Mill's *System of Logic* is in the MIT Archives. Pencilled in the top right-hand corner of the title page is "WBR Univ. of Virginia 1847."

14. Alexander Dallas Bache, *Report on Education in Europe to the Trustees of the Girard College for Orphans* (Philadelphia, 1836), 546–561, quoted by Frederick B. Artz, *The Development of Technical Education in France, 1500–1850* (Cambridge, 1966), 237.

15. Artz, *Development of Technical Education*, 155, 237, 266.

16. Giorgio de Santillana, "Positivism and the Technocratic Ideal in the Nineteenth Century," in *Studies and Essays in the History of Science and Learning Offered in Homage to George Sarton*, ed. M. F. Ashley Montagu (New York, 1946), 247.

17. Rogers, *Life and Letters*, Vol. 1, 15, 26.

18. Ibid., 244, 245.

19. Comte, *Positive Philosophy*, 30.

20. "Constitution of the American Association for the Promotion of Social Science," Article II, printed in the Proceedings of the Association, 1865–66. The printed text is in the Boston Public Library, call number 5561.5.

21. Ibid, 6–12; Rogers, *Life and Letters*, Vol. 1, 91, 371.

22. John Wright Buckham and George Malcolm Stratton, *George Holmes Howison, Philosopher and Teacher* (Berkeley, CA, 1934), 48–49, 59, 60.

23. "A Report on the Behavioral Sciences at the University of Chicago" (University of Chicago, 1954), 5, 6

24. National Academy of Sciences, *Engineering Education and Practice in the United States: Foundations of Our Techno-Economic Future* (Washington, DC, 1985); Report to the Faculty, Committee on Undergraduate Admissions and Financial Aid, MIT, May 1989.

25. Gertrud Lenzer, *Auguste Comte and Positivism: The Essential Writings* (New York, 1975), 14; Herbert A. Simon, *Models of My Life* (New York, 1991), 36–47; John Maynard Keynes, *Two Memoirs* (London, 1949), 78–103.

26. Edwin H. Land, "Generation of Greatness: The Idea of a University in an Age of Science," Arthur Dehon Little Memorial Lecture, MIT, May 22, 1957.

Technology and the Liberal Arts
Ann F. Friedlaender

Introduction and Overview

In *The Machine in the Garden,* Leo Marx argues that American intellectual history has been characterized by a tension between two competing myths, one "pastoral" and the other "progressive."[1] In the late 17th and early 18th centuries, the self-sufficient, independent farmer embodied pastoralism, while the inventors, "mechanics," and industrialists who were creating European industrial society were the avatars of progressivism. According to Marx, the new world attracted settlers imbued with one or the other of these ideals. The "pastoralists" sought to escape from the increasingly complex and hierarchical world of Europe and hoped to establish a society in which individuals could live in simple harmony with nature. The "progressivists" saw in the vast open lands of the new world endless opportunities to harness nature in the service of industrial development.

The Jeffersonian vision of a self-sufficient, rural, agricultural society with little dependence on European technology was essentially pastoral. In contrast, Hamilton's vision of an America founded on a strong, technologically based manufacturing sector that would exploit the continent's vast resources was progressive. In the event, the progressive ideal has dominated the American economy and social structure, while the pastoral ideal has become a mainstay of American intellectual life.

Henry Thoreau's move to Walden Pond and Thorton Burgess's move to a cottage on Cape Cod were manifestations of a desire frequently expressed by 19th-century American intellectuals to create a simple, pastoral life that minimized their dependence on "the machine." Over and over again, as Marx notes, 19th-century and early 20th-century literature expresses respect for the pastoral and the rural and distaste for the urban and the technical. This ideal takes a variety of forms, from the steamboat that runs over Huck and Jim's raft to the demonic vision of the factory within the *Pequod* that turns mighty whales into sperm oil and other manufactured goods. Among 20th-century writers, Hemingway's vision remains essentially pastoral, as does Faulkner's. And it is not a coincidence that Gatsby's world collapses with an auto accident in the "valley of ashes," a creation of a technical and mechanistic society.

While the pastoral ideal has become less conspicuous in post–World War II literature, it is still very much a part of American culture. Manifestations include advertising images such as the Marlboro man and, on a more serious level, those environmentalists and antinuclear activists who argue that "small is beautiful" and, in the extreme case, that technology is equivalent to pollution and the desecration of the planet. In these groups, the pastoral ideal has become not only a basic element of culture but also a political force.

I believe that the pastoral ideal has also been a dominant force in American education, and that our inability to create a reasonable reconciliation of the pastoral and the progressive myths has not only contributed to the relatively narrow base of American liberal education, but has also been the source of the continuing distrust that laypeople have for science and technology.

Technology is not going to go away. One need only consider the technical revolutions being wrought by the silicon chip and modern biology to recognize that, like it or not, science and technology are going to assume an even more central role in American society in the future. Thus it is essential that we confront the educational implications of the continuing tension between these two visions of American society, recognizing at last that the introduction of the machine need not necessarily involve the destruction of the garden. I am going to argue that it is time to blur the line between a liberal and a technical education so that we can create technologists and humanists who are conversant and comfortable with each other's world. To do so, however, will require a fundamental restructuring of the educational paradigm.

A Liberal Education for the Future

While all predictions are inherently suspect, it seems clear that for the near future American society will continue to be technologically based. Given the importance to society of scientific and technical developments, the educated layperson cannot afford to abrogate responsibility for the management and implementation of science and technology. Taking up this responsibility, however, requires some understanding of basic scientific and technological principles. Thus a liberal education for the future must equip individuals with sufficient background knowledge to deal with current science and technology and to understand changes that will occur in the future. Such an education should address not only the fundamental physical and natural orders but also how society deals with their manipulation—whether it be through genetic engineering, the smashing or fusion of the atom, or the storage of information in a ceramic chip.

Thus a liberal education for the future should integrate humanistic and societal concerns with those of science and technology to permit individuals to move freely between the "scientific" and "humanistic" cultures. By giving individuals the ability to explain the natural, social, and cultural orders through a variety of perspectives, a liberal education should also create an ability to integrate different forms of knowledge and cognitive processes.

A liberal education should be broad and should incorporate essential elements from the arts, humanities, social sciences, natural sciences, physical sciences, and engineering sciences. It should give students an understanding of the social, cultural, and humanistic forces that have shaped society and also of the scientific and technological revolutions that have constantly affected humanistic and social endeavors. Finally, it must give students a mastery of the different types of cognitive processes that are required for understanding different disciplines. Only with such a mastery can individuals move freely and comfortably among different areas of knowledge and thus achieve the ability to connect and reassemble knowledge from these areas in creative and imaginative ways.

A liberal education is also an education in scale, permitting individuals to handle many levels of hierarchy, to feel as comfortable with phenomena at the level of the atom as at the level of the individual or of society, or even at that of the galaxy or the universe. Only by learning to move through these hierarchies of being and organization can one learn to recognize where organization and structure may have special meaning and to make meaningful connections between the structure of molecules, for example, and that of the societies that use the molecules in a particular fashion.

At the same time, a liberal education must also be an education in depth and permit individuals to master a given set of disciplines. Only with such mastery can one gain the confidence required to make connections among the various disciplines and fields and undertake the integrative process. Thus specialization is an integral part of a liberal education.

The essential element of my ideal liberal education for the future is the integration of science and technology with the traditional liberal arts. This implies, specifically, that all students learn the basic principles of matter: how fundamental forces are shaped into atoms, molecules, and cells, and how these in turn are shaped into tangible matter and its physical manifestations—whether it be nuclear energy, supercomputers, or genetically engineered cows. Without such an understanding, the broad mass of citizens will continue to view the creations of technology as marvels and and will approach them with a certain fear and awe—indeed as 20th-century analogues to the railroad that seemed to many critics to intrude upon and desecrate the American pastoral landscape and society. Without such an integration, the technological xenophobia implicit in the pastoral myth will continue to dominate America's liberal education to the detriment of society.

A Technical Education for the Future

The problem of breaking down artificial barriers does not lie exclusively with the liberally educated upholders of the pastoral myth. Just as current liberal education tends to foster a skeptical, if not disdainful, attitude to science and technology, the converse is also true in that modern technical/scientific education leads students to question the benefits and relevance of the liberal arts. I will stress two distinct but equally important contributions that the liberal arts can make to technical education: (1) They can give students a broad base for understanding the social, political, and ethical consequences of science and technology; and (2) they can offer exposure to a wide range of approaches to problem solving and conceptualization that will enhance students' creativity and imagination.

Science, Technology, and Society

It is increasingly important that scientists and engineers not only understand the limitations of the progressive myth, and consequently see the need for a wise and sensitive use of technology, but also be able to deal with a citizenry imbued with the pastoral myth, who are consequently skeptical and fearful of the impact of technology on society. To illustrate the problem, I will look at the issues of nuclear energy and the Strategic Defense Initiative.

The recent experience of the nuclear energy industry in the United States is a good example of the consequences of unresolved conflicts between the progressive and pastoral myths. Indeed, I believe that it clearly illustrates the need for "progressive" technologists to have a broad understanding of the social, ethical, and political consequences of technology as well as the need for the "pastorally" educated layperson to understand basic scientific and physical principles. To the technologists, nuclear power was a difficult engineering problem that, if solved, could offer a safe, economical source of energy. While they clearly underestimated the technical difficulties and thus the cost of nuclear energy, they also underestimated the public's visceral fear of anything nuclear and its reaction to the very real, if extremely low, probability of a catastrophic accident.

The accident at Three Mile Island and the disaster at Chernobyl have probably placed a burden on the nuclear industry from which it cannot recover; but even in the absence of these massive failures of technology, it is clear that nuclear engineers failed to assess and address the negative externalities—both physical and psychological—generated by nuclear energy. Many would not even admit that the topic was political and should therefore be open to democratic debate. The result was that groups of concerned citizens took it into their own hands to create litigation and legislation that delayed and escalated the costs of constructing new nuclear facilities throughout the United States. Thus the problem was transformed from the "progressive"/technical one of developing plants to produce nuclear energy at reasonable cost at some acceptable level of risk, into the "pastoral"/social one of creating a no-risk environment. While it is unclear that a rational debate could ever have developed with respect to the first issue, given society's fear of nuclear explosions, it is also clear that by ignoring the social context within which the second debate was evolving, the engineers lost whatever initiative they might have had to lead that debate. Although it is doubtful that a more socially and politically aware generation of engineers could have changed the outcome of this debate, it is likely that the existing generation lost it before it ever began.

Although the relaxation of the cold war has made much of the Star Wars debate academic, it is still instructive since it has many of the same characteristics as the controversy over nuclear energy. On one side is an anxious and concerned citizenry who doubt its feasibility and also question whether, if it worked, it would increase national security. On the other hand are the scientists and technologists who view it as a challenging intellectual problem, with little concern for its social implications. Indicative of this attitude is the comment by Roger Guillermin, a noted physiologist, that "Science deals with the acquisition of new knowledge. The use, including the misuse or ill use, of that knowledge is the realm of politicians" and, by inference, the concerned citizenry.[2]

More revealing, perhaps, was an article in the *New York Times Magazine* a few years ago about the Star Wars initiative and the work of a scientist named Peter Hagelstein.[3] Hagelstein came to the Livermore Labs with a desire to develop X-ray lasers for medical purposes, but within the environment of the lab, the scientific problem of "pumping" or increasing the power of X-ray lasers by an order of magnitude became dominant in his mind, and the question of whether it was "pumped" by a nuclear bomb or by a laboratory laser became immaterial. As it happened, in a moment of creative insight, he provided the key to nuclear pumping on which the Star Wars technology is based, and the focus of his work at Livermore was consequently changed from biology and medicine to weaponry.

Peter Hagelstein does not seem to be someone who is insensitive to the implications of his research—he is clearly troubled by the arms race—but the following quote reflects an attitude that is probably shared by many scientists: "Until 1980 or so I didn't want to have anything to do with nuclear anything. Back in those days, I thought that there was something fundamentally evil about weapons. Now I see it as an interesting physics problem."[4] (Hagelstein recently left Livermore and is now teaching at MIT and doing "pure" scientific research.)

And so we are brought back to Roger Guillermin's statement about the role of the scientist—to solve difficult problems and add to the sum of knowledge, but not to worry about the consequences of this knowledge. While art for art's sake is certainly a workable ideology, I would argue that scientists can never afford to be oblivious to the consequences of their work. This does not mean that they should stop their inquiry, but it does mean that they must take more responsibility to ensure that society uses their findings wisely and responsibly.

I, for one, am not sufficiently technically informed to know whether nuclear energy is a "good buy" or what is the appropriate level of national defense—whether the question is that of MX missiles or the implementation of a Strategic Defense Initiative. I do know, however, that society can only lose if scientists and technologists imbued with the progressive myth refuse to enter into the debate—saying, in effect, "Trust us. The issue is too complex for you, the layperson, to understand."—and if, in response, the concerned and pastorally inclined citizenry closes ranks against all science and technology.

If scientists and engineers are to understand the full range of implication of their discoveries, they must be able to understand society's reaction to these discoveries. This means that they must understand (1) the cost of and nature of the market for new technical developments; (2) the way consumers and societies as a whole assess the risks associated with new developments; (3) the political process and the role of lobbies and pressure groups; (4) ethics; and (5) the sociology and psychology of individuals and groups who are confronted with new technological advances. The increased level of technical and scientific education that I have advocated for all laypeople would also help reduce the existing level of mutual distrust. While knowledge would not necessarily lead to a meeting of minds, it would certainly increase the likelihood of rational debate and decision making, to everybody's benefit.

Creativity and Design
Let me now turn to the question of the role of the liberal arts in creativity and design. While relatively little is known about the creative process or the determinants of imagination, there is considerable evidence that truly creative scientists and engineers are not only broadly educated but have strong parallel interests in the arts and humanities.[5] For example, Robert Fulton and Samuel F. B. Morse were considered important painters in their day, George Washington Carver and James Clerk Maxwell were accomplished poets, and Albert Einstein and Max Planck were skilled musicians. And to return to Peter Hagelstein, it is worth noting that he plays and composes for the violin and viola.

Although existing evidence is largely anecdotal, it does suggest that creativity and imagination are developed and enhanced by an exposure to many approaches to knowledge and ways of knowing. In this connection, the following quote from an article by Jerome Wiesner is as relevant today as it was when he wrote it some 25 years ago:

> The basic habit of maintaining a skeptical, actively critical point of view toward all knowledge and opinion must be ingrained. Such habits will minimize the development of unconscious inhibitions against consideration of the widest possible spectrum of possibilities in approaching a problem of interest.
> There must be encouragement and stimulation of imaginative and unconventional interpretations of experience in general; this is particularly true in problem-solving activities. It is important, especially in childhood and early youth, that novel ideas and unconventional patterns of action should be more widely tolerated, not criticized too soon and too often.
> Since interrelated bodies of factual material can be more efficiently scanned and searched, the accumulation of facts during training should emphasize principles, laws, and structural relationships.
> We must explicitly encourage the development of habits and skills in looking for, and using, analogies, similes, and metaphors to juxtapose, readily, facts and ideas that might not at first appear to be interrelated. Early in this development we should foster a clear understanding of the special character and usefulness of the private, informal process of conducting a search for new ideas and insights; and of the distinction between it and the equally necessary but more elaborate and rigorous machinery needed for verification of results and their systematic development for incorporation into the accepted body of knowledge.[6]

These are skills generally taught to undergraduates not in the scientific and engineering disciplines but in the disciplines of literature, history, writing, music, and the other liberal arts. For this reason, the education of engineers and scientists should include a large dose of humanistic inquiry that would expose them to the intellectual and conceptual processes of these disciplines. Only by such an exposure can individuals understand the correlative aspects of the various disciplines and free their minds from the shackles of narrowly conceived, purely technical approaches to problem solving.

Conversely, liberally educated individuals would greatly benefit from understanding the processes of hypothesis formulation and testing in science that are needed to translate intuitive and creative insights into parts of an accepted body of knowledge. The same rigorous procedure can often be applied usefully to nonscientific creative insights. While intuition can start the process, its execution is typically mundane and often involves analytical processes that are closely akin to those employed by the scientist or the engineer.

Conclusion

Given the technological basis of modern American culture and society, we can no longer afford to live as two societies, one pastoral and the other progressive. The costs of the implicit conflicts between technological and humanistic cultures are simply too high. Thus an education for the future must ensure that the lines between the sciences and technology are blurred with those of the arts and humanities to ensure that both technically educated and liberally educated individuals have a basic understanding of and appreciation for each other's world.

In practical terms, this means that the distinction between a liberal education and a technical education should be diminished, if not eliminated. A liberal education must include a strong exposure to the sciences and engineering and include at a minimum mathematics, physics, chemistry, the life sciences, and computer science. Conversely, a technical education must include a strong exposure to the humanities, arts, and social sciences. Moreover, efforts should be made throughout the educational process to integrate the humanistic and scientific worlds and concepts so that a "technically" educated scientist or engineer is as comfortable with the world of arts and letters as a "liberally" educated humanist is with the world of science and technology. Thus we might want to think of "liberal technology" as an educational paradigm of the future in place of liberal arts.

Victor Weisskopf, one of MIT's most distinguished physicists, has argued that, ideally, there should be no absolute distinction between the sciences and the humanities, between usefulness and playfulness, indeed, between the much heralded two cultures.[7] But if all are part of a whole, are simply different manifestations of a single, unifying process or culture, they are certainly not treated or recognized as such. Thus the primary goal of education—technical or liberal—should be to make these similarities apparent and thus to enhance one's creative abilities, regardless of whether one ultimately becomes a scientist, a technologist, a humanist, or even an academic administrator.

Let me close, then, by affirming my belief that there is indeed only one culture, but many different ways of knowing and of acquiring knowledge. As we, as individuals, become knowledgeable about these various realms of inquiry, our abilities to make the connections will grow. Thus, in a truly successful education, the many cultures should merge into one.

Notes

1. Leo Marx, *The Machine in The Garden: Technology and the Pastoral Ideal in America* (New York: Oxford University Press, 1964).

2. Quoted in Robert Scott Root-Bernstein, "Creative Process as a Unifying Theme of Human Cultures," *Daedalus*, Vol. 113, no. 3 (Summer, 1984), 198.

3. William J. Broad, "The Secret Behind Star Wars," *New York Times Magazine*, Vol. 134 (August 11, 1985), 32.

4. Ibid., 49.

5. This section draws heavily on Root-Bernstein, "Creative Process."

6. Ibid., 213.

7. Ibid., 203.

Mary, Theresa, and Elizabeth
Cynthia Griffin Wolff

MIT, which likes to think that it is little more than a loose amalgamation of productive units (immensely creative near-chaos), has certain deep, inflexible structures that astonished me when I first arrived. For example, there is a pecking order here that defies all the appearances of casual disarray, a highly elaborated, delicately calibrated scale of status that is honored everywhere in the Institute. It is based not on salary or rank or even on one's standing in the world at large, but only on intelligence rather narrowly construed.

"H has hardly published a word," I overheard someone say of a senior administrator one day at lunch. "Such a shitload of respect and so little on paper. Who does he think he is, anyway?"

"A heavy hitter," his companion replied, tapping his head solemnly with his middle finger as he tried to negotiate a tuna sandwich, "a BIG BRAIN, that's who he is."

"But literal-minded," persisted the first. "When you talk to him, you have to touch every base—in the *right order*. He's a *linear thinker*," he concluded triumphantly, "a *LINEAR THINKER!*"

This was a serious hit. His companion stopped chewing and paused reflectively. "True. He is not an abstract structures person. And he is not imaginative, can't *do* lateral thinking. But have you ever seen him handle numbers? It's a pleasure to watch him do budget—the sheer . . . virtuosity . . . yes, the virtuosity of his mind manipulating figures. Like watching a professional athlete. Speed and precision and style. All right, maybe not an absolutely *first-class brain*, but still, a heavy hitter. Not the *very* best, but right up there— right up there. . . ." The two men nodded in unison and went on with their lunch, happily in agreement.

When I first heard a conversation like this, I did not pay much attention. It was an aberration, I thought, an encapsulated instance of eccentricity. But then I heard more of such exchanges—and still more. And finally (the ultimate compliment), I began to be invited to join in them. "This is crazy," I thought, "this obsessive assessment of the mechanical workings of other people's mental processes. This public preoccupation with the most private elements of thought—always construed in a competitive mode. It is like . . . peeing against a wall." The metaphor that had sprung so quickly and aptly to mind was itself a revelation, though not a happy one. "I have joined the faculty of an Institution that is conducting a non-stop, world-class peeing against the wall competition!" Good grief.

Now I have begun this little account of Mary, Theresa, and Elizabeth (who are three of my undergraduate students) to show that I know the rules of the Big Game here. And also to give me an excuse for saying, right up front, that each one of these young women is (according to the rules of the Great Game) a potential "heavy hitter." They are very, *very* intelligent, my three students—precisely as gifted as any of their male counterparts. But their concerns and their lives may be different from what has seemed "usual" for so many years at MIT. That is why it is important for you to get to know them. Just be sure to remember, as you read along, how immensely gifted each one of them is. This is the last time I will remind you.

I met Theresa first—in the fall of 1985, her freshman year. In the dingy atmosphere of my classroom (MWF, 8:30—Introduction to Fiction), she was luminous. In part it was a function of her appearance. Skin that was tawny and flawlessly beautiful, blue-black hair that fell to her waist in soft, graceful waves—gestures that had the effortless grace of a dancer. So even if she had said nothing very interesting or insightful, it would still have been a pleasure to watch her move and speak. But the merely *visual* turned out to be no more than a bonus. Her capacity to understand not only particular novels, but the more general properties of fictional worlds (in the abstract) was phenomenal. "In *Bleak House*," she observed during class discussion one morning, "distances are not constant. At least the distance between London and Bleak House is not a constant . . . . and that is very strange. Does it have anything to do with what you told us at the beginning . . . about the opening paragraphs having verbals, but not finite verbs—not sentences postulated in *time*?" This was to have been the burden of my own remarks that day (and I had asked the students to ponder this issue of distance in the novel before they came to class that day). Of course I had not expected anyone to cover so much intellectual terrain in such short order. It was not the usual case.

"What is your major?" This was almost my first question when she came to office hours. (And why are you at MIT instead of at one of the great liberal arts universities? This I *did not* say. But I thought it.)

"Chemistry. Organic." She looked down. "It's so *beautiful* . . . almost like a unique museum, but then always fluid and changing. And I love it. Sometimes it's like I can walk around *inside* the compounds and just look at how they work. My father . . . ," she paused with apparent embarrassment. "My father is not an educated man. But he is very proud of me, even though I am a girl, and in my family, girls are . . . ," she shrugged, "not important like their brothers. But I was very good at math—that is why I got to come up here. I scored very high on the standard math tests, and the nuns sent the results to the Archon Foundation (which promotes the advancement of talented students on the island), and they gave me a scholarship to Exeter—and even then, I had to have special tutoring in my junior and senior years because I had completed all of their advanced placement math courses. So my *father*, who has a great respect for mathematics, wanted me to major in that. He doesn't understand about chemistry. But when he realizes that I can get a great job—" she looked up, tossed her hair back and laughed, "well, *then*, even my father will like chemistry."

By the next term, when Theresa took my American fiction course, she had acquired a steady "friend." Bradford was tall and blond and awkward with language (at least in my presence). When she came to my office hours, he occasionally came with her. Sometimes she teased him in an affectionate way, and it seemed to make him come alive. They were very fond of each other—very kind to each other— and it was pleasant that April and May to watch them together. I had hoped to keep seeing Theresa, but little by little she disappeared, absorbed (I supposed) by the demands of the chemistry she loved so much.

Two years later I met Mary and Elizabeth in the same course (Introduction to American Literature: TTh, 1:30–3:00). They could not have been more different.

Elizabeth was tall and blond; she came from downstate Illinois, and she seems to have known all her life that she wanted to become a doctor. "My grandfather was a minister," she told me. "But he was a homeopathic healer, too. Do you know what that is?" I nodded. "Well, when he stopped practicing, he gave me his bag. Of course I'm going to become a conventional doctor, so the bag was just symbolic, passing down a tradition. And it was lovely, more like an elegant jewelry case—or I should say a jewelry *cabinet*—than a doctor's bag. Because it's big, almost too big to carry around. Fitted out, you know, in such a beautiful, orderly way. Dozens and dozens of little bottles with labels that say things like 'arsenic' and 'arnica.' And when he gave it to me, he explained that in *his* kind of medicine, he administered very powerful elements and compounds, sometimes even poisons, in very small and very pure doses. That all the *art* of his work was in learning to recognize the exact nature of his patients' bodies—how each uniquely worked—as well as in being able to recognize their illnesses. So my grandfather would spend hours, sometimes, listening to a sick person describe his pain and his symptoms—and his life in general, too. And then he would prescribe little adjustments and wait to see their effect—tinkering and listening until his patients got better. And they *did* get better. They almost always got *all better*. When Grandpa's eyesight forced him to stop practicing, his patients begged him to train somebody himself to take care of them. Of course he couldn't do that. But he saved his bag for me."

Elizabeth was strangely silent in class. She wrote five papers, all superb, and she was animated enough when we talked alone. But in class, there was this impenetrable reticence. I never understood that. Once her whole face lit up with glee. I had asked her whether she was going home for the summer. "Oh yes! I can't wait. I *hate* the city."

"Do you have a lab job there?" I wondered.

"Yes, and of course I also tassle. Mom says it's crazy to keep that up. It's too much to do along with my lab job, and it tires me. But I love it. So I still do that, too."

My face must have betrayed my bewilderment. I hoped it did not also betray the wildly lurid images that I had conjured of this sedate young woman's aestival activities. "You . . . *tassle?*" I ventured. She burst out laughing.

"Near where we live they experiment with hybrid corn. They grow three rows of each variety—many, many rows to a field. You have to cut off the tassles or they cross-breed, and that would ruin the experiment. All the high-school kids make summer money cutting off the tassles—tassling. Everybody at home uses the term, and I just forget that people here have never heard of it." And she broke out in giggles at my wide-eyed ignorance.

Elizabeth's classmate, Mary, could not have been more different. She was the most energetically cheerful student I have ever had.

Intelligently talkative in class. And eager. For example, I often refer to books that are not on the students' reading list. From time to time, I will amplify the reference. "This is a book that every educated person ought to read. Buy it now and keep it. Eventually, you may find the time, and then you will already own it." Not every student will copy down the citation. Some never do. But Mary made a note of every one. What's more, she often stayed after class to ask for the exact citation of a book that I had mentioned only casually (and whose title I had not written on the board).

Mary came into my office hours three times that term. She had gotten A's on her papers, and needed no instruction about the best ways to develop an argument—although she always began by asking respectfully about this matter. She just wanted to discuss the reading a little more. The first time, she was fascinated by American Puritanism (a reaction that is virtually unheard of among undergraduates these days); in particular, she wanted to ask about free will and predestination. If God already *knows* everything, does that mean He has *planned* it? How can we have free will if God already knows our future?

The next time, we had read *Benito Cereno* and *Billy Budd*, and she wanted to ask about spontaneity, innocence, and justice. "My grandfather and grandmother were interned during the Second World War," she remarked casually during the course of the discussion.

"How awful," I said.

"Yes, my mother was born right after they got out. They were very angry, and for a long time they would not talk about it at all. Sometimes I think that's why my mother had so many children (Mary had six brothers and sisters). Just this conviction that something terrible could happen again, and you have to have a lot of kids to make sure one or two survive. And of course, my grandparents hated America." As soon as this last came out, she looked guilty. "Probably I shouldn't be saying this to a professor of American literature. But I *love* America. I went to Japan when I was 16. I have cousins there. Do you know that half of their marriages are still *arranged*—and that if a woman has a serious job or teaches college, everyone thinks she is peculiar and she is . . . well, 'shunned'? America is not like that because in American there is progress, so I am glad that my family decided to stay—even though my father's brothers have been urging him to come back."

In her third visit (she had come to ask why anyone had EVER taken *Daisy Miller* seriously), Mary told me that she had decided to major in Chemical Engineering. "It's reality-oriented, practical," she explained, "Not just some airy theory—numbers on the page and hypothetical particles" (she made a face). "It's the thermal properties of fluid moving through a pipe—real nuts and bolts. You figure out how to make thinks work right—how to make them work *now*, in the real world, under real world conditions. It's almost like some terrific computer game except it's actually there and you can check it and keep changing the game until you win. Anybody would love that! And there are women in the field. It's not like the math department where the only women around are post-docs. This is important to me because I'm not going to get married and I'm not going to have children, so I want my career to be really good—really successful. Satisfying. Because I'm looking forward to my *career* to make *me* happy. My brothers and sisters will have enough grandchildren for my mother and father, and I can be this rich "auntie" who takes them to concerts and helps them through college."

Her vision of the future was so clear and persuasive that for a very brief moment, the good-natured 18-year-old face seemed to become the face of a benign, successful middle-aged woman—the chemical engineer and "auntie" that Mary was determined to become.

Mary and Elizabeth slipped out of sight after that term. Occasionally I would see them in the halls, and we would nod or say hello. But they (like Theresa, I supposed) had been absorbed by the "real" world of MIT—chemistry, biology, chemical engineering. It was both a surprise and a pleasure, then, to find last term (fall 1990) that both had enrolled in my small, upper-lever course entitled "The Literature of Suffrage and Abolition" (TTh, 3:30–5:00). But the biggest surprise was Theresa, who was also enrolled in the course. She should have graduated more than a year ago, and she was only now (and only by over-loading) hoping to graduate the following June. The young women I had known as freshmen had all changed. Life is never simple.

After the first meeting of the class, Theresa stayed behind to talk with me. "What's up?" I asked right away. "Did you take time off? I thought you had already graduated."

"Why don't you walk with me for a while. I have to pick up Charity, and I can explain things."

"Charity?" (I wondered whether this was some bizarre mode of referring to her scholarship money.)

"Charity is my little girl. She is two and half." I looked quickly at her left hand: no ring. Of course she caught my furtive glance and understood my curiosity (actually, I later realized that this must have been a regular occurrence), and she bridled. "No, I'm *not* married. Bradford's mother and father did not like the idea of a Puer-r-r-rto R-R-R-R-Rican daughter-in-law. But did they like the idea of a grandchild! Oh, yes! Too much. They say I'm an unfit mother, you see, because I am determined to finish school. Because I want to have a career. You see, when they found out I was pregnant, they told Br-r-r-r-radford no marriage. No way. But at the same time, they told him to demand custody. Can you believe the crust of these people? And he caved in to them! Can you imagine. But he got nothing in the end. Nothing! Who would want him, anyway—a man who would do that to us! His name isn't even on the birth certificate."

"And you named her 'Charity'?"

"You think every little Puerto Rican girl has to be named 'Rosalita'?"

"No, but why *that* name? It seems to have . . . connotations. Won't she wonder when she gets older?"

"It was my grandmother's name. A *family* name. *My* family. And now we are a family, Charity and I." (Theresa spends $640 dollars a month for childcare at MIT. The TCC [Technology Children's Center] is "affiliated" with the Institute, but independently run; it does not accept children under the age of two years, nine months. The *personnel* at the Center make every effort to be helpful [Theresa emphasized this repeatedly]. There are listings of private individuals who are licensed to provide daycare for younger children, and when Theresa first returned to school at the Institute, she interviewed a number of such people until she found the right person—"just a nice, OK person," she said; "it wasn't easy. And of course it costs a fortune—all of it. But what should I do? I did not want to have an abortion. I didn't have religious reasons. You know, general objections left over from the nuns. I believe very very strongly in every woman's right to do what she wants about this; and for those women who want abortions—I *fight* for their right to control their bodies. I *fight* for the RIGHT to abortion. But that was not *my* choice. Can you understand that? *I* did not want to abort my baby. And shouldn't I have *that* choice? And now someone has to support us—and that someone has to be me. Of course Brad and I used 'precaution.' Ha! C—ter W—lace maybe ought to do better quality-control. What do you think? One ruptured rubber and look at my life!")

"Are you still majoring in organic chemistry?" I asked. "Are you still planning to go to graduate school?" Theresa looked away and said nothing at all for a moment.

"Yes, I'm finishing the chem. degree. But now ... Now I'm going to spend the summer working and getting education credits so I can teach high school. How could I possibly go on to graduate school—now? Maybe later ... if they would take me back. There aren't any tenured women with children in the chemistry department. So I don't fool myself." She sighed. "Just one thing at a time, these days. I think my life will seem different when Charity is a little older. But you know—*Bradford* is already in graduate school. What do you think of that? It's the thing I'm absolutely the *most* resentful about. One broken rubber, one broken career. Mine."

The subject that these three women were taking (along with seven others—three more women and four men) was a course in the implicit narrative structures of the literature of the suffrage and abolition movements in 19th-century America. We read some works of fiction (*Uncle Tom's Cabin* and *The Yellow Wallpaper* among others), but we read essays too—by Higginson, Stanton, Douglass; and we read slave narratives. Inevitably, we covered a lot of history, but always (since this was a literature course) I kept trying to bring my students back to the exercise of identifying the implicit, underlying abstract narrative structures that informed these movements. This involved first dissecting the "standard plots" (for example, what was the "story" about what women were "supposed to be" that had led this nation to deny them the vote—to deny married women any rights to property or even to the guardianship of their own children?) Next, we examined the implicit narrative assumptions of the reformers who wanted to change things—because until this nation can create *workable new* stories to replace the ones that have excluded women and blacks from the mainstream of American life, the hard-core discrimination against them will remain unchanged. (In order to change what we *do*, we must first change the patterns that shape our thinking about these things.)

During the first two-thirds of the course, the students read the assigned texts, and we discussed them together. Then we had a two-session exam on the reading. The rest of the term was devoted to individual presentations of research projects, and this was by far the most interesting part of the course.

Elizabeth's topic—childbirth and birth control in 19th-century America—became the focus for a great deal of discussion. Before the American Revolution, she explained, childbirth had been an affair entirely conducted by women: the woman giving birth, the midwife, and other women in the family (or in the neighborhood) who came in to help. "Sometimes there was almost a party atmosphere; they took care of both the mother and baby. And if it was a *first* baby, they taught the mother a lot about how to care for the child. Then when male physicians began to take over the birthing process, the other women were excluded; the pregnant woman was isolated—and mortality *increased!* Mortality for both the mother and the neonate increased! The techniques became much more aggressive and invasive and controlling."

Elizabeth had been working in Professor I's biology lab for more than a year; he had been immensely encouraging and kind, and he felt strongly that she ought to go into a research career ("look how many *more* people you might help with your discoveries," he had argued). Yet she still clung to the hope of practicing medicine and was now deeply divided about her future. This indictment of the profession she had idolized was very painful.

"Well, what story lies behind the change?" another student inquired. "Was it just money, money, money? The greedy doctors (even back then) looking out for the chance to expand their practice and make another buck? I mean, like they say here at the Institute, 'if it ain't broke, don't fix it.' Midwifery wasn't 'broke,' was it?"

Elizabeth shook her head slowly. "No. Although *some* things were improved when the doctors took over. Surgical procedures were devised to correct fistulas . . . and . . . some lives were saved, perhaps, by forceps deliveries. . . . But no, in MIT terms, midwifery wasn't 'broke,' and now in the 1990s, the medical profession is finally going *back* to the use of midwives!" She paused, seemed about to speak again, and then fell silent.

This presentation was not easy assignment for the young woman whose grandfather had been such a thoughtful and noninvasive homeopathic healer. Nor did it become any easier when she moved on to her discussion of the Comstock Law (1875) and America's consistent opposition to any form of effective birth control. We talked for several classes about the tone of the language used in this 19th-century debate—surprisingly violent and hostile to women's sexuality. One of the male students quoted from his own research on attitudes concerning sexuality: "William Acton [a 19th-century British physician who published texts on 'women's' illnesses and was powerfully influential in both England and America] said that a 'modest woman seldom desires any sexual gratification for herself. She submits to her husband's embraces, but principally to gratify him; and, were it not for the desire of maternity, would far rather be relieved from his attentions.' If people really believed *that*, then maybe they would assume that birth control would eliminate sexual relations altogether."

"No," Theresa said; "I think it's just that men are terrified of women's bodies (which seem complicated and peculiarly powerful), and they need to control us and control our decisions about having children."

"But things have changed now," Mary objected. "In America there is always progress. Look at MIT. They dispense free condoms. No questions asked. Now everybody says women are deviant if they *don't* want to have sex. If, in fact, they don't act just like men—beer blasts and all-nighters."

"What do you think the underlying narrative is here and now at MIT about women and sexuality—birth control and childbirth?" I asked. This was a question that none of the students had thought to raise. Safely ensconced in their discussions of the 19th century, they were nervous about bringing the issue up-to-date. And so for several minutes (which seems like a very long time in the classroom) there was complete silence while they sat and thought.

Finally Elizabeth spoke out. "It's all right for the women of MIT to have sex. I guess in fact it's all right for the women here to engage in just about any form of sexuality that they find appealing—so long as it is mutual and voluntary and not exploitative. And so long as they don't get pregnant. It isn't *sex* that is taboo at MIT. It's pregnancy."

"Oh, it's even all right to get pregnant," Theresa interjected derisively. "You just have to be willing to get an abortion. You know. Quick. Clean. No after-effects. Back to work without missing a day." And she tossed her head with visible contempt.

"Well, babies are not MIT's problem," Mary said.

"Babies are *everybody's* problem." Elizabeth replied quietly, but with an authority and determination that would have seemed impossible only three years earlier.

"More and more I can't understand why everybody just thinks that they can shove the whole responsibility onto the mother. You're always talking about how good America is . . . what a wonderful country. Well, in case you don't know. . . . America has an appalling rate of infant mortality. Our politicians seem to care immensely about the *abstract issue* of 'children'—of 'babies.' Many of them feel nothing so strongly as the inalienable rights of *unborn* babies, fetuses. And so they would outlaw abortion. They talk with such righteousness about abortion! And yet once the baby is *born* ...," Elizabeth paused. Briefly, her eyes threatened to fill with tears. But she steadied herself, regained her composure, and went on.

"America is a *democracy*—we are supposed to be providing for our future *citizens*, here. All these babies will become our citizens and our voters.

"So I can't understand why Americans, who ought to care the most, have so little regard for real babies, *real* children. Can it perhaps be just hostility against *women*? That seems so crazy. Yet I listen to the actual words people use. Like, 'She got herself pregnant.' I am majoring in biology, and I am here to tell you that if you can *get yourself pregnant*, you can make medical history. Babies are not just the *mother's* 'problem.' No civilized country—no viable society of any kind—has *ever* supposed that a baby could be just the 'mother's problem'!"

One of the male students had been listening to this outburst with increasing discomfort. "We're not all such bad guys. Look at Lotus Corp., right down the street in Kendall Square. They have state-of-the-art childcare for their employees for just a nominal fee, about $100 a month. I think that company's run by MIT grads. And hey, Elizabeth, it's for sure an *American* company. So we're not *all* bad. Besides, we can't solve America's problems here in some humanities course at MIT. So lay off."

Everybody lapsed into an embarrassed silence. They were such basically nice young people, and of course they were not in the least angry with *each other*. And now, nobody quite knew how to dispel the awkwardness. Finally Elizabeth broke the silence. "Of course you're not all bad guys. In fact, the guys that are here in this room are unusually *good* guys (I mean, look—you're taking this course). And I'm sorry if I made you feel defensive about America's big problems. I guess this is just something I have to think about if I want to go into the practice of medicine. And it helps, it actually helps, to talk about it."

At this point, Elizabeth was prepared to conclude the discussion. But Theresa was not. "Okay. Forget America and its very big problem with hypocrisy toward women (and maybe even more toward *minorities*—because bear in mind how many of those dying America babies are children of *color*). Let's just stick to the original question. Are babies 'MIT's problem'?

"MIT went out and deliberately looked for women students—

encouraged them to come—gave them scholarships. But MIT hasn't changed very much except the raw numbers of female students. The Institution hasn't changed its fundamental attitudes, and it hasn't really changed its support services. MIT wanted more women *if* the women would really pretend to be men in disguise."

"Yes. For example, the MIT pace and pressure are especially terrible for women's bodies," Elizabeth added, drawn briefly back into the discussion. "My female friends get sick from 'all-nighters' and their periods become irregular. What's wrong with MIT shows up more readily in its effect on women's health because their endocrine systems are more complicated. And I think the Institute just thinks, 'That's too bad. Women just aren't tough enough.'"

"Yeah," said the forthcoming male student. "That's probably true. And then, men do tough it out, and then they drop dead of a heart attack in their fifties. Big advantage."

"But I am *certain* MIT will get better," Mary insisted. They just need time. After all, MIT is committed to progress." . . .

None of the other presentations prompted discussions that were as heated or uncomfortable as this had been. Nonetheless, my own personal sympathies were perhaps most powerfully moved by a different dilemma—one that seemed entirely to baffle Mary's usual cheerful optimism. She had done a report on 19th-century women in business. As a kind of epilogue, she decided to take a look at women in chemical engineering today—the "business" in which she was herself so determined to succeed. As it happened, the sophomore enrollment in chemical engineering that year had shown such a dramatic skewing by gender (of 66 students, 44 were women and only 22 were men), that the fate of women in the field—the effect of this apparent *stampede* of women into the field—was already under investigation by others at the Institute. And the initial results were sobering. Mary presented them to the class in a voice that suggested (to me—perhaps not to the others) a profound sense of betrayal. Disbelief and bafflement, but beneath that, betrayal.

"Women who are already in the field have mixed reactions to the 'feminization' of chemical engineering," she said. "On the one hand, they are pleased and proud to have been in the vanguard—to have forged the way. But on the other hand . . . . Well, on the other hand, they worry that as *women* move into the field—no matter how good they are—there will be less prestige and lower incomes."

"For the same work?" a male student asked.

"Yes. I mean that's the peculiar thing. For the same work. Maybe even for better work. Women come into the profession; men flee; and salaries and status go down." She shook her head. "Maybe it's because people think that when *women* have a job, it is just a 'second income'—not important." She shook her head again. "But for *some* women, of course, it will be their entire life. They will have given up a good deal for that career. And even so, maybe just because they are women, no matter how intelligent or how good they are at their work or even how hard they work, they can *never* be as successful as the men. Never be a CEO. Just because they are women. No matter how much they work, no matter what they do. . . ."

A week or so before I started to write this essay, I was having dinner with a friend from the history department.

"Ah, you're writing an essay for the Vest volume, Cynthia. What are you going to say? Something arcane and deep about Emily Dickinson? Something witty and urbane about Edith Wharton's satires?" (My friend—a very dear friend—has a sardonic sense of humor and an arch manner, and he likes to put me on).

"No, I'm going to write about my students, my female students and their particular difficulties at MIT. Their identities will all be suitably disguised, of course. But the essential stories will be true."

"Beating the ancient and honorable drum for the women's movement, Cynthia?"

"Not exactly, my friend. Don't be so sarcastic; you might even learn something. . . .

"We have terrific students here at MIT. I've been teaching for 30 years, and I have never encountered more wonderful young people. Brilliant and surprisingly generous. And for a brief while, they are entrusted to us. We teach them many things during the four years they are here. Science and engineering (and other incidental things like humanities) are, perhaps, the *least* important. We teach them coping skills, for example. And yet the coping skills that MIT inculcates are dysfunctional in the long run! For example, this is the most stressful institution I have ever seen. Not *useful, exhilarating* stress. Not creative. Just stress for its own sake. Like basic training. And this is what we present to our students as a model for *life!* Permanent, life-time basic training. And so they get mid-career burn-out! Big surprise.

"But most important, we give them—or we *ought* to give them—a sense of 'community.' They get their first lesson away from home right here. Their first real 'work' experience, too. Yet even the grown-ups fail this test: we are so busy coping with stress and *competing* with each other that we have almost no sense of mutual responsibility—no sense of community. Under these circumstances, how can we ever help our students to create the communities of the future? And if they cannot create viable communities, nothing else matters. It all goes up in smoke.

"These problems with MIT as an institution—as something that shapes people's lives and shapes America's future—show up most clearly in its treatment of women. So that's what I want to write about. This fundamental failure of values. This destruction of what might otherwise be very great here."

My friend The Historian poked at his oyster for a moment in silence, speared it, and chewed ruminatively. Finally he said, "Did you hear Paul Gray's Inaugural Address?"

"No. Either I had just arrived and didn't attend. Or I arrived the following fall. Anyway, I didn't hear it. Why do you ask?"

"Well, Gray made this eloquent speech about doing something to alter the pace and orientation of MIT. He talked about all those indisputably good things, Cynthia. You know. Family values, more time for thought, for creativity. He may even have mentioned a commitment to the welfare of the larger community; I don't remember. Anyway, it was a very fine speech." The Historian paused to demolish another oyster. "There were two standard-issue MIT types sitting behind me. I have no idea who they were and it doesn't really matter—just the generic model, you know. And when Gray finished his speech, one turned to the other and muttered, 'Just let him *try* to change this place.'" He sat back beaming. Not pleased. Just sardonic—jaded and very, very weary. And then he continued.

"MIT has a very big plug into the money/power line, Cynthia. They *like* what they're doing. In fact, they're convinced that they're doing *exactly the right things*. You can write your earnest little essay. It will probably be therapeutic for you. But it won't change MIT. Go ahead and beat your little drum. This place is already marching full speed ahead to the music of a great big brass band!"

Addictions and Recovery at MIT
Eve Odiorne Sullivan

MIT's mission, according to Samuel Jay Keyser's "Agenda for the Next Decade," is to create a community of scholars in which each contributes as much as possible to the increase of human knowledge. He also calls for MIT to devote some of its resources—time, energy, and money—to the study of social disorganization, and he also suggests that the Institute take a greater leadership role in promoting economic competitiveness and productivity. In this essay I will argue that the most serious drain on our resources, the most common cause of social disorganization, and the greatest barrier to productivity is addiction. I will attempt to define addiction and will present a view of recovery as a community effort that can lead us to some simple, positive changes. Changes in Institute policy and procedure can, I believe, create a climate at MIT that will—like a breath of fresh air in a stuffy room—give us new energy for the endeavors of mind and hand that are the Institute's primary mission.

I will describe various addictions and suggest how these support each other, both in the individual and in groups. I recognize that recovery can take more than one path. The best-known path follows the 12 steps of Alcoholics Anonymous, but there are other rational, secular organizations that have the same end: support for sobriety.

What is addiction? Addiction is an obsessive behavior that can involve substances, activities, or relationships that can be extraordinary or mundane—food as well as drugs, work as well as risk-taking, aggressive, or violent activity, and relationships in the family or at work. My thesis is that a common pattern exists in all addictions: Some aspect of an individual's life is out of control. As the individual attempts to control that substance, behavior, or relationship, the effort results only in increasing frustration and ultimately in desperation. Statistics on the toll of addictive disease could fill several pages, but these few on violence—the last resort for control—will suffice. The June 1991 *Harvard Mental Health Letter* states, "Violence is both a major public health problem and a serious concern for mental health professionals. One and a half million aggravated assaults and a half million incidents of child abuse are reported each year. . . . Much of this violence, including a third of all homicides and up to 75 percent of assaults, occurs within families or between lovers."

A Chinese proverb reveals the essence of addiction: "The man takes the drink. The drink takes the drink. The drink takes the man." The activity—involving an ingestible substance such as food or drink, an activity such as work or play, a relationship either emotional or physical—is at first undertaken by choice. I am hot and thirsty and choose to drink. That one drink tastes good, and I have another. Finally, and this process may occur on a small scale over the course of an evening as well as on a large scale over a period of months or years, the activity overtakes me. At that point I can no longer choose *not* to engage in the activity.

Most of us see tobacco and alcohol use as addictive or, at the very least, as a risk for addiction. No one who has had a family member or close friend die from lung cancer or in an alcohol-related car crash will argue against strong measures to control the use of these two substances. But we must also learn to see that ordinary daily activities—both intimate and public—and even highly valued endeavors such as scientific research can become addictive. Addictions from compulsive hand washing to obsessive sex are grist for the mill of talk shows and popular magazines. Gambling is now commonly viewed and treated as an addictive disease. Indebting oneself with money or time—compulsive shopping and compulsive working—have the same effects in one's life as any other addiction and respond to some of the same treatment. Any smoker who has quit smoking only to gain weight knows that addiction can take many forms.

Is is too far a leap to a point of view that sees the pursuit of excellence and its dark side, perfectionism, in academic work as similarly addictive? I think not. MIT takes justifiable pride in its accomplishments and reputation. Scholarship is, as Charles Vest noted in his inaugural address, "simultaneously an individual and a communal activity." A great risk comes when we lose the trust that others in our larger community, our nation, and our world have placed in us.

President Vest quoted Pogo's famous line about the enemy being us. Where is this enemy? Naming individuals is certainly counterproductive. It would serve no purpose to seek out the 10 percent of our population who suffer the disease of chemical dependency and require these individuals to wear a scarlet letter labeling them Alcoholic or Addict. They are—some drinking, some sober—present in all groups in our community: faculty, students, and staff. Some of them may already have sought help, some may be ready to, and some may be mired in denial, hearing only the inner lies: "I can handle it. It's not that big a problem."

Better to ask, Where is the enemy in each of us? And better to spend our energy looking for and changing the attitudes and behaviors that directly contribute to the disease and those that passively allow it to flourish. What are these attitudes?

"Work hard, play hard" acts as a prescription for drunkenness when "play hard" becomes "party hearty," which in practice means drink to excess. "Pursuit of excellence" may become intolerance of error, which may in turn engender fear and deceit. Recently, EECS coursework had to be checked for identical code because students, under pressure to get it in and get it right, copied work.

Concern for confidentiality on the part of medical staff may translate into lack of support for a supervisor with an active alcoholic employee. The employee is allowed early retirement, and the problem of alcoholism is never addressed. Professional staff's concern for the autonomy of a young person in trouble with alcohol and other drugs undermines parents' authority and supports the disease with the attitude that "We have to wait until he is ready to accept treatment."

If we wait and do little or nothing until someone asks for treatment, we end up playing along with the disease. You can call it disease, or dysfunction, or cowardice, or stupidity. Whatever it is, it encourages us to be silent when we should speak up. Someone we work with, or work for, or live with, or a friend we care about, may be in trouble. We see little signs and ignore them. We may be in some difficulty ourselves and decide not to speak up and ask for help. But what happens when disease is left untreated, when problems are ignored?

Like other illnesses, addictive diseases may be primary, chronic, progressive, or terminal. Unfortunately, professional treatment usually arrives on the scene—like paramedics in an ambulance—after the disease has progressed for some time and a serious crisis has occurred. It is much better for me if I can recognize my own "disease of choice"—or if someone confronts me with care and concern—and I seek help at an early stage. The best care is preventive care. In the case of addictive disease, both on a personal and community level, prevention means time and attention given to our own self-care and to decision making. Professional help in the form of disciplinary measures may be called for and are most useful as leverage to force individuals to confront their problems.

We have named the four qualities of addictive disease. Let us now look at the four stages of addiction. In the first, I experience a mood swing from whatever the activity is. I feel good smoking, drinking, eating, working, arguing, speeding, shopping, achieving sexual satisfaction, or whatever. In the second stage, I seek a mood swing from the activity and still feel good or satisfied some of the time. In the third stage, I am preoccupied with the activity but experience a good feeling, a "high," less and less often. In the fourth stage, I seek but do not find good feelings, yet I keep on doing whatever it is, becoming ever more desperate.

I would suggest that only when we acknowledge the similarity among all addictive diseases and confront our own complicity at various levels in the addictive system can we make the changes necessary to free ourselves, our companions, and our communities from this plague. How can we do this? How do we confront "the enemy" face to face? The key word is *confront*. Caring confrontation among members of a community is essential if the individuals and the community are to become well and stay well.

Another phrase useful in describing the process of confrontation is *controlled crisis*. Rather than wait for disease to progress, we can take the initiative and make an early intervention for recovery. This must be done in the context of a community. For any social unit, small or large—couple, nuclear or extended family, classroom, workplace, or the whole nation—to support sobriety in its members, it must become a therapeutic community. What makes a therapeutic community work is individuals openly confronting failings in themselves and others. To the extent that MIT does not promote caring confrontation among the members of its community, it fails to support our health and achievement. In failing to confront individuals who are in trouble, we lose that most precious and intangible element of our lives: our peace of mind and heart.

I am calling for MIT to inaugurate a new era of support for sobriety. The policy and procedure implications of this commitment are both simple and basic. We must acknowledge that denial is often an indicator of addictive disease. "I can stop anytime I want" or "It's not a problem in my work group or living group" may well mean the opposite. The dishonesty that has been so much in the headlines and that eats away at the foundations of scientific integrity is accompanied by similar denial. Building, or perhaps rebuilding, a community in which people are willing to acknowledge their own mistakes and to confront others on theirs is essential if the Institute is to remain a leader in the national and international scene. I would like to suggest that MIT undertake a leadership role in this difficult and important area of social concern.

What will this commitment mean in practical terms? It can include any of the following. It should include them all:

•An Institute Committee on Addictions and Recovery.

•Addiction/recovery training in the mandatory Safety Office presentations and in all new-hire orientations at every level of the Institute community, including personnel officers, medical staff, support staff, and physical plant staff, as well as faculty and administrators.

•Annual presentations to the Academic Council and to administrative officers, departments, residence/living group/fraternity advisors, student groups, staff working groups, and unions.

•Annual workshops on substance abuse, to be given on three consecutive days, either at lunch or on paid work time, for faculty, administrators, personnel officers, support staff, and physical plant workers. These would include first-person stories, such as those given in the May 1991 Women's Forum, along with role-play of workplace and family intervention scenarios.

•An Institute Colloquium, "How To Be Sober, and What to Do When Somebody Isn't," presented after discussions such as those described above have been held in all the various groups within the larger Institute community; similar at Residence/Orientation Week and Parents Weekend.

•Establish one or more "Wellness Dorms" with specific programs of support for sobriety and healthy living.

•Incorporate in all presentations issues related to food addiction. Also include the issue of harassment (sexual and other), since this can be viewed through the addictions/recovery lens as an obsession in which a behavior, person, or relationship acts as a drug.

MIT has been known since its founding over a century ago for excellence in intellectual and practical endeavors. Indeed, the icon of MIT is *mens et manus,* mind and hand. This essay proposes that a third element, spirit, is essential to the success of endeavors of both mind and hand. This element was both assumed and present in the era of MIT's founding, but in our time it is at the very least besieged and at worst entirely missing. Like a stool with only two firm legs and the other leg weak or broken, our society at large and the Institute as a microcosm are at risk. Each of the three basic spheres of our experience—affective, cognitive, and behavioral—is equally important. We ignore the spiritual and emotional sphere at our peril.

The risk inherent in inattention to or rejection of inner life is individual. We lose touch on a daily basis with the excitement of learning along with the frustration of failure. As we ignore the smaller daily ups and downs, we miss the the joy and sadness that attend all of life's passages: the end of a school year, the beginning of a friendship, birth and death.

The risks attendant to ignoring inner life are also communal. Without a sense of spiritual connection to one another, we have no source of the energy necessary to be present in one another's lives. Without a sense of connection to our family members and our extended family—our classmates and workmates—the Institute family, we have no energy or willingness to confront one another with honest differences. As Walter Massey, director of the National Science Foundation, said in this year's commencement address, "We have moved away from small . . . communities in which misconduct is easily observed, toward large anonymous . . . enterprises in which tasks are fragmented and accountability is hard to ascribe." It is more difficult to stay honest and easier to cheat in a large community. But we can keep each other honest if we stay honest ourselves, on an emotional level. Addiction is at the core a disease of the feelings. I must confront my own negative feelings, fear, anger, and resentment, and work them through if I want to have energy and ability to experience fully the good feelings that my life and work offer.

"Sorrow carves out so that joy may fill up," a college friend wrote, and her words have come back to me many times. I wish you, President Vest, your family, and MIT's large family and many friends, immense joy in your individual achievements. And I share the joy and pride in our common work. At the same time, I wish for you and all of us the spiritual strength to acknowledge sorrow and pain. In order to break the addictive cycle, we must go back into the pain, but this time come out a different way, one that does not dull our hearts and minds. MIT also suffers. But we do not have to suffer in silence. We can talk about it, together.

A Personal View of Education
Alan V. Oppenheim

Education has been an important part of my life as a child, as a student, as a parent, and as a faculty member at MIT. For me, the experience of learning new things and of imparting knowledge to others carries with it a special sense of excitement and gratification. Although I have never had any formal training as an educator, my career seems to suggest that I have an intuitive sense for it, and certainly I have had a lot of experience at it. I have also learned a lot about education from my wife, who, besides having a career in education, has in my view excellent instincts as a teacher and a parent.

My professional life revolves around teaching and research, which at MIT are symbiotic activities. I was once asked which one I would give up if I had to choose between them. I hope never to have to make that choice because the two are complementary and each is exciting and fun. However, if forced, I think I would choose teaching over research. I particularly enjoy the thrill of a successful classroom performance and of working closely with students. I am also awed by the enormous leverage and responsibility that teaching provides in terms of propagating one's knowledge, standards, and ideals. Many of my former students are now teachers, as are many of their students. Teaching, like parenting, influences an endless succession of generations.

In this essay I would like to express some of my personal views about teaching and learning. There are, of course, many equally effective styles of teaching and learning, and I present my thoughts on the subject in the modest hope that they will cause others to think about what works best for them.

There are obviously many facets to learning and to our roles as educators. An important part of learning involves a straightforward absorption of basic facts and skills. In learning arithmetic, we memorize multiplication tables and practice procedures for long division. In language learning, we work on expanding vocabulary and conjugating verbs. In history, we learn dates, events, places, and their interrelationships. At another level, however, there is a need to develop the judgment, insight, and intuition to use these skills and facts appropriately and creatively.

I vividly recall my own experience when I was first learning geometry in high school. I was intrigued by the fact that one could start with a few simple axioms and definitions, use them to prove some simple theorems, then prove more complex theorems from those first simple ones, and so on. As I was learning the basic skills of geometry, I developed the impression that geometry and other branches of mathematics are carried out by simply "turning the crank," that is, by routinely applying previous theorems to generate new ones, in the way that a machine might do it. It was not until some time later that I came to realize that a major part of creativity in mathematics is in knowing what theorem to prove, in having the deep insight required to conjecture that something previously unknown is in fact true and to guess at why it is true.

How one learns or teaches creativity and insight is not at all clear to me. A prerequisite is clearly a fluency with the basic skills associated with a subject. Much of education, at least as it is done today in the United States, focuses on working problems that are well formulated and easily associated with a particular part of a textbook or lesson. In a typical arithmetic or mathematics class, for example, students may be asked to prove a particular theorem, given the previous theorems in the textbook. The students know in advance that solving the problem requires only the material they have covered up to that point. In working with graduate students who are learning to do creative research, one of the teacher's most difficult challenges is to break this pattern. In research or industry, the problems to be solved are much less structured and typically do not fit into nice compartments. Structuring problems in such a way that we know which tools to use is often the hard part. As I often comment to my research students, "Once you get an issue to the point where it looks like a problem on a problem set, the hard part is done."

I would suggest that we need to start in the earlier grades, exposing students to unstructured problem solving by means of problems for which there is no clean solution available. Of course, this needs to be done in a balanced way. In the United States, the pendulum has swung in several directions over the years. In the 1960s, there was a trend toward open classrooms and what was referred to as "exploratory learning." This approach was often overdone to the exclusion of training in basic skills, and the result was often students who could not spell or compute. The pendulum then swung strongly back toward the basics. In the 1980s, there was—as there still is—considerable emphasis on trying to achieve a balance between exploratory learning and the development of the basic skills.

The distinction between basic skills and creativity brings me to the issue of the role of computers in education, particularly at the grade school and high school levels. As an analogy, I recall recently watching a tennis class. On some of the courts a student was on one side of the net with a machine on the other side. The machine shot tennis balls across the net with different speeds and directions so that the student could practice the skills required for returning a variety of shots. The machine was consistent and tireless. Other students were working closely with a coach to develop instincts, anticipation, and strategies and to get feedback on form, technique, etc. In my view, the computer is particularly effective in a role similar to that of the tennis ball machine. It is excellent for endless and often entertaining practice with drills to learn and improve basic skills. For this it is often much better than a "live" teacher, since it is infinitely patient, nonjudgmental, and very forgiving. However, I believe that the individual teacher or "coach" and personal interaction are needed to guide the student in developing judgment and creativity. Put another way, the computer is an excellent tool for teaching, as are textbooks, slide projectors, chalkboards, and workbooks. But it cannot teach creativity and insight, other than that which is naturally gained through drills and experimentation. When used effectively, however, it relieves the teacher of a variety of routine tasks. In geometry, for example, it can help the student learn definitions, do practice problems at varying levels of difficulty, explore relationships, and gain precision with the skills and experience with the concepts. In history, it can help in learning facts and their associations and can potentially offer elegant ways, such as hypertext, to access a large store of information.

Computers enrich the educational environment in a variety of other ways, particularly at the university level. The availability of campuswide networking and electronic mail greatly enhances communication at all levels. The ability to simulate mathematical expressions and engineering systems offers an opportunity for "theoretical experimentation" in many courses to develop insight and to help in visualization. It can do all these things in an entertaining way that can make the experience fun and exciting for the student.

I feel strongly that having fun is very important. I am reminded of a conversation in which I asked a research colleague why he was working on a problem that to me seemed somewhat irrelevant. His answer, literally, was that "the whole idea is to have fun." That phrase has stayed with me, and I repeat it often to my students. If, as teachers, we can impart to our students a positive feeling about themselves and a sense of fun and gratification in learning, then many other desirable aspects of their education will follow naturally.

Of course, I find that students need to be reminded constantly that along with the fun comes a lot of hard work. As Thomas Edison said, invention is 1 percent inspiration and 99 percent perspiration. As another way to put it, I realized many years ago when doing a major project on my home that you can paint 90 percent of a room in 10 percent of the time. The walls are easy. It is careful attention to the trim and other fine details that makes the difference. This is consistent with my earlier comment that along with the development of exploratory and creative skills, we cannot lose sight of the importance of precision with basic skills.

I have occasionally seen it suggested that computers may eventually take over the educational process, and these suggestions usually cite various dramatic experiments and results. It seems to me that these experimental results are often due to what has become known as the Hawthorne effect. Many educational experiments that are highly successful in pilot programs are considerably less successful when installed on a larger scale. This results from the fact that it is often not the details of the experiment that make it a success, but simply the enthusiasm and excitement that are transmitted to the students from the teachers, associated with the fact that they are trying a new approach. In this sense, it is often worthwhile to introduce and experiment with new techniques and educational tools (such as computers) strictly for the novelty they offer to students and teachers. In addition, of course, these techniques and tools may well have long-term value, as I believe computers do when used for the right purposes.

I am a great believer in the importance of the teacher's role as mentor. As teachers, we are, in effect, intellectual parents. There are many styles of mentoring that are appropriate and effective, just as there are many styles of parenting. All that is important is that the style be comfortable and natural for both student and teacher.

Some of my colleagues prefer a relatively formal student/teacher relationship. My personal style is to encourage students to interact with me informally, starting with their use of my first name. Particularly with my graduate research students, I often strive to develop a relationship in which we become friends and colleagues. I must say that in my role as a parent at home also, I often say to my two children that I hope that they see me not only as their parent but also as their friend.

In thinking about the dual roles of teacher and friend, I am reminded of a situation many years ago involving one of my former doctoral students, now a successful scientist. We were both learning the sport of windsurfing. On one particular morning, we met to review his thesis progress, and in fairly authoritative terms as his teacher, I emphasized the importance of his working harder and making quicker progress. As it turned out, that afternoon conditions were particularly good for windsurfing, and I was in somewhat of a dilemma as to whether, as his teacher, to encourage him to spend the afternoon working or, as his windsurfing friend and "student," to encourage him to take advantage of the good weather. You can guess how it turned out. Even now he and I still fondly recall that day and dilemma. In general, I feel strongly that it is appropriate and meaningful for students to feel that I am both their teacher and their friend. Of course, in situations in which there is conflict between the roles of teacher and friend, the responsibilities of the teacher should always come first.

It is also often the case with students that there are times of role reversal in which they are the teacher and I am the student. I believe that this role reversal is as important to me as it is to them. This happens frequently at MIT during Independent Activities Period, when essentially the only requirement for a course being offered is that there be someone interested in teaching it and someone interested in taking it. During IAP many intriguing courses are taught by students with special skills and interests and taken by faculty and other students. Two of the most enjoyable courses that I have taken during IAP were "close-up magic" (card tricks), and "lock-picking." Both were taught by former classroom students of mine. They were quite frankly thrilled with the role reversal, and so was I.

Beyond this simple role reversal, all of us are and should always be students as well as teachers. In fact, in the Electrical Engineering and Computer Science Department, faculty are encouraged to take occasional courses as students. We attend all the lectures, do all the required assignments, and take the exams alongside the regular students. I have done this several times. The most recent was an undergraduate core subject taken mainly by freshmen and sophomores. For most of the semester I sat next to one particular freshman who clearly grasped the material much more quickly than I did and to whom I often turned for help. One day after class he asked me why I was taking the course. When I told him it was because I would be teaching it the next semester, he looked totally astonished and then burst out in laughter. I can say from personal experience that periodically returning to the classroom as a student is both exhilarating and humbling. Among other things, it gives us an excellent view of ourselves from the students' perspective, thus providing an opportunity to heighten our sensitivities as teachers. The opportunity for faculty to take occasional courses has been further enhanced and formalized recently in my department through the establishment of the Adler Scholar Program in memory of Professor Richard Adler.

With regard to classroom interaction, again there are many styles that are appropriate and effective. I personally try to be highly interactive in my lectures, even with large classes. Interestingly, when I am a "student" in a class in which the teacher is highly interactive and looks for participation from the class, I tend to be nervous and to feel intimidated when asked to respond on material that I am just learning. However, I also find myself paying especially close attention to the lecturer, so that if I am called on, I have at least a reasonable chance of having a meaningful response.

As I have tried to stress, I view my role as a teacher to involve not only imparting skills in a specific area but also developing students in a broader sense. Of particular importance is the development of creativity, insight, and a positive, confident self-image. Confidence and self-esteem are essential since creative processes inevitably involve disappointment and failure, set off, one hopes, by occasional successes. In guiding doctoral students through a major research project, I consider it important to push them to their creative limits and to instill in them the self-confidence to carry out creative research on their own. As we all know, in order to fly, it is important to believe that you can.

Self-confidence and self-esteem are, of course, personal qualities that are best instilled by both parents and teachers at an early age. This point was made to me very emphatically by my wife, Phyllis, when our daughter was about to enter school. She was tested for early admission and passed easily. As a proud parent, I was pleased that she was ahead of her age in intellectual ability, and I was in favor of her entering school early. Phyllis clearly saw it as preferable that our daughter wait. As she pointed out, in making such a decision, it was important for us to consider the physical as well as the intellectual aspects of development. The younger children in a class are often unable to do things that the older children can, simply because they are a few months behind in motor skill or behavioral development. Nevertheless, the younger children often mistakenly see this as reflecting a lack of innate ability, and it affects their self-confidence. For this reason, it is generally better to be one of the older children in a class than one of the younger ones. Having watched my two children and others progress through school, I have no doubt that Phyllis is absolutely right on this point.

There are a few additional points that I am eager to convey to my students at some stage in their education. The first is the importance of visualization or, equivalently, planning ahead. I remember as extremely significant a moment in my freshman year in college when a close friend asked me to describe how I would most like my career to turn out. He encouraged me to express my wildest, most optimistic dream. With the disclaimer that I could not imagine this ever happening, I said that I would most like to know a subject well enough to write a book on it, to be a professor at a top university, and to lead a small, successful research group. While there are many elements of good fortune that contributed to the fulfillment of this dream, a significant element was the fact that I was encouraged by my friend to articulate and therefore visualize it. I now try to take each of my research students, and any other students with whom I have a close association, through the same process of visualization with regard to their careers.

Another point is the importance of having a set of serious interests that can offer refreshing diversions from work. My own outside interests revolve around sports. A sometimes frustrating but also exciting aspect of creativity is that it seems to happen at a subconscious level and rises to the surface only when ready. In my experience, it is often the case that, after laying the proper intellectual groundwork, I can encourage creative inspiration best by doing something completely different that puts my mind in a relaxed, somewhat meditative state. It is literally true that the creative breakthrough on my doctoral thesis occurred for me two-thirds of the way up a ski lift in Vermont.

I often say to my students that no news is bad news and that all news is good news. The phrase as I have worded it is just confusing enough that it encourages them to stop and think for a while. What I mean specifically is the following: One of the exciting things about a creative endeavor such as research is never knowing exactly where you are going to end up or how you are going to get there. Along the way there are many surprises, some of which may on the surface seem disappointing. For example, perhaps a student learns that their great "discovery" has been discovered before, or is wrong. My view is that nothing is ever completely new and that all news is good news—that is, there is often as much or more to be learned from the disappointments as from the successes. In fact, a very important role for teachers is to show students how to recover and learn from mistakes, failures, and disappointments.

I would like to conclude with one more mental image that I am eager to convey to my students. In 1980, I gave a series of lectures at Tsinghua University in Beijing. My host asked me at some point how I pick research problems to work on. I tried to explain the importance to me of choosing problems based on my instincts and on the fun and excitement associated with an uncertain outcome. Because of the language difficulties, I eventually turned to a metaphor that I have since used quite frequently. As I explained it to my host, choosing a problem to work on is like standing at the edge of a field of tall grass, from which there are several tails sticking out. The idea is to tug lightly on each to get some feel for what might be at the other end. For example, is it a mouse or an elephant? Eventually, instinct and judgment will suggest which tail to follow, and it should then be pursued vigorously for the fun and excitement of discovering what it is attached to.

Although I have been teaching for many years, this is the first time that I have had the opportunity to express my personal views about education in a somewhat formal way. When I started to put this essay together, I was not quite sure whether my train of thought would lead to a mouse or an elephant. I hope that I have at least conveyed the sense of personal pleasure that I derive from teaching and the importance that I place on the relationship between student and teacher. I was particularly fortunate to have had a number of excellent teachers who had significant impact on me personally. Among them were Amar Bose, who showed me by example the importance of exceptional standards, and Ben Gold, who showed me by example the importance of friendship and informality. I hope that in a similar way I can convey these values to my students. I particularly look forward to many years of being a friend and teacher to my students and to being their student as well.

Note

This essay is adapted from a talk prepared for presentation to the Kankaju School Systems, Osaka, Japan.

Leadership through Science and Technology
Michael L. Dertouzos

The Need for Change

Science and technology are still regarded as separable commodities from the rest of human activity. Worldwide, the common wisdom on this attitude goes something like this: If you are a leader of an organization or a nation, set your goals, your local or national priorities, and then shop for whatever science and technology you need, almost like buying potatoes, to get the job done!

This mindset has kept scientists and technologists in the role of servants to wiser people who, based on "the best that has been thought and said in the world," have been making the important decisions. Whether in business or in government, this attitude, which was accepted on all sides and worked well while science and technology played a relatively small part in human affairs, is no longer viable for several reasons.

First, with global peace within closer reach than ever before, with the emergence of the Pacific Basin and Europe as formidable industrial powers, and with the discovery of free markets by the Soviet Bloc countries, and probably soon by China, the world is poised for an economically vibrant 21st century—one that, we hope, will increase the well-being of many more of our fellow human beings. The accompanying generation and consumption of ever more sophisticated products and services will create, among other requirements, a need for ever more sophisticated technology and for more science.

Second, and quite independent from this shifting economic demand, progress in science and technology seems to be moving briskly, perhaps even accelerating along what seems to be a historically inevitable trajectory that builds on a rich foundation of past advances and a healthy combination of creativity and serendipity. Three major areas—the life sciences, materials science, and information science and technology—are expected to have a profound impact on our lives during the 21st century. Surely, there will be others as well, whose potential influence is not yet as easily visible.

Third, it is clear that nations and people are increasingly confronted with massively complex *sociotechnical* issues, such as the legal, ethical, and practical consequences of new achievements in the life sciences; the balancing of energy needs with the requirements of human health and safety; the balancing of development needs with environmental protection; and the possibility of massive violations of privacy stemming from electronic information transactions.

Expressions of opinion on these profound issues abound. What is nearly absent in current debate, however, is an informed assessment of the interdisciplinary interactions needed to deal in a realistic way with these complex societal issues. For example, an objective forecast of the future of the environment would require, among other things, a painstaking assessment of the consequences of all of the world's major industrial activities, augmented by the even more difficult assessment of the impact of expected industrial changes.

An increasingly skeptical world blames science and technology for these complex problems. And with good reason. In the absence of technology, none of these ills would be present. But neither would any of the benefits that technology has brought! If the world wants to enjoy cheap energy, rapid transportation, sophisticated products, affordable food, and electronic recreation, and there is ample evidence that it does, it must try to understand and control the ways in which science and technology interact with and cause interactions among individuals, society, and nature.

To deal with the creations of science and technology requires a more technically literate world population—from boardroom to assembly line, from manufacturers to providers of services, from local to federal governments, and from primary and secondary schools to liberal arts and technical universities. The necessary educational changes go well beyond a need for more science and technology. In manufacturing, for example, rigid systems of production in which workers, suppliers, and equipment are dedicated to the single purpose of keeping the gigantic wheel of production turning, no matter what, are on their way out. They are being replaced by flexible approaches in which people and technology interact tightly; there is a creative balance of teamwork and individualism, with an overall focus on productivity; workers are more involved, better rewarded, and continuously educated; and narrow specialization no longer brings the advantages it once did. In *Made in America*, the report of the MIT Commission on Industrial Productivity, we summarized the educational changes needed to meet this new challenge as follows: "MIT should broaden its educational approach in the sciences, in technology, and in the humanities and should educate students to be more sensitive to productivity, to practical problems, to teamwork, and to the cultures, institutions, and business practices of other countries."[1]

In light of this increasing role of science and technology, it is unrealistic to expect that tomorrow's leaders, whether in business, government, or academia, will be able to choose a wise course without a thorough and deep understanding of the technical and societal dimensions of their decisions, as well as of the associated interactions among those dimensions.

In a nutshell, the growing and inextricable blending of science and technology with human activity calls for a shift in the strategy of technological institutions such as MIT.

A New MIT Strategy

MIT seems to follow no explicit strategy. Yet, I believe that most of us have been functioning under an implicit strategy, which I would characterize as *leadership in science and technology*, recently tempered by a vague push toward increasing educational breadth. This push toward a more integrated education is also felt by predominantly humanistic institutions, many of which are sprinkling their curricula with a healthier dose of science and technology.

At MIT, one hears these days spirited discussions polarized around those who want well-rounded students versus those who prefer students with sharp edges. To the former, the latter are too narrow for tomorrow's world. To the latter, the former are too broad and in risk of dilettantism. To my thinking, neither the add-and-mix nor the stay-as-we-are solution will work as well as a different, bolder approach that recognizes the changing roles of science and technology. I believe that MIT and the world's technical universities should set as their new strategic goal: *leadership through science and technology*. This small change in wording from the implicit strategy of leading *in* science and technology to the explicit strategy of leading *through* science and technology packs a wealth of consequences, which I will try to describe specifically in terms of MIT.

Changes in Instruction

The premise that science and technology will play a greater role and will blend further with human activity means that at least a fraction of tomorrow's nontechnologists, such as lawyers, business people, doctors, educators, and politicians, will need to have a considerably more solid grounding in science and technology.

On the other side of the artificial humanist-technologist divide, it also means that some fraction of tomorrow's technologists will need to go beyond their narrow technical specialties. In *Made in America*, we said that this new cadre of MIT students should be characterized by "(1) interest in, and knowledge of, real problems and their societal, economic, and political context; (2) an ability to function effectively as members of a team creating new products, processes, and systems; (3) an ability to operate effectively beyond the confines of a single discipline; and (4) an integration of a deep understanding of science and technology with practical knowledge, a hands-on orientation, and experimental skills and insight."

To prepare tomorrow's leaders for a world in which science and technology are better integrated with human activity, we need to think now about how to educate both technologists who wish to be broader and nontechnologists who wish to have a stronger background in science and technology. Toward this goal, imagine a future in which all of our undergraduates go through a broad four-year core curriculum that is richer in basic scientific, technological, managerial, and humanistic foundations, at the expense of some courses that today provide disciplinary specialization. Let us call this, for the moment, a technological alternative to a liberal arts education.

Nontechnologists graduating from this broader curriculum would continue their studies in law, management, medicine, diplomacy, and so forth. They would differ from their counterparts who enter these institutions with degrees from traditional liberal arts colleges by their strong technical education, and they would probably be dubbed *tech-based lawyers*, *tech-based doctors*, and *tech-based managers*.

The other side of this coin, the new cadre of MIT technologists, would, upon graduation from this four-year program, continue their studies at MIT or elsewhere for one or two years to gain a professional degree in their specialty, most probably in a field of engineering. This lengthening of the time it takes to earn a professional degree from four to five or even six years would reflect our intent to provide greater breadth while preserving depth and avoiding the dilettantism that accompanies shallow generalism.

Against the criticism that lengthening the educational term is expensive and unnecessary, I offer the following question: Why should a technologist expect to lead a complex world after four years of study, when a doctor and a lawyer must go to school for seven or eight years? (There was a time when they, too, went to school for only four years.)

By initiating this change, we would, in effect, redefine what a broader-based technologist should be and we would exert worldwide leadership in educating this new breed.

At the same time, I expect that a sizable fraction of our faculty and students will continue to focus on narrower technical and research-oriented paths, as many of us do today. And this leads me to an important admonition: The continuation of our leadership in science and technology, in the narrow sense, is crucial to our future. Besides being the starting point for our new strategy and besides being what we know and are known to do best, it is also the essential prerequisite for carrying out the new strategy. For if we lose our proven ability to explore the inner secrets of science and technology, we shall be unable to lead effectively through science and technology. People trust and listen to advice only if it is backed by deep knowledge and experience. Such has been our past strength and such it must continue to be. We must therefore avoid any temptation to produce a species of generalist, discipline-independent experts.

Accordingly, the new strategy of leading through science and technology should not be viewed as replacing but rather as requiring and building upon our proven strategy of leading in science and technology. In due time the mix of our students and faculty among these three classes of technical specialists, broader technologists, and tech-based nontechnologists would find its own equilibrium.

The educational reform that would make these changes possible should not be viewed narrowly as impinging only on the technology side of MIT. This broadening and deepening involve not only a lot of what we teach today but also the interactions among these subjects. This is the reason why it will be neither intellectually adequate nor efficient to pursue the new strategy by simply forming additive collages drawn from the subjects we currently teach. Our educational leadership will most likely hinge on the way we go beyond this approach and integrate instruction with research, as we have always done, distilling essential but mutually reinforcing concepts into fundamental core-curriculum offerings.

Changes in Research

In research, leadership through science and technology would mean a greater involvement with practical problems and with sociotechnical problems. The case for a greater participation in practical problems stems in part from the changes discussed earlier and in part from our history. MIT helped pioneer the post–World War II transformation of engineering from a collection of cookbook procedures to a science. This was a result of the observation that the key contributions in wartime research were made by physicists and mathematicians rather than by engineers. Unfortunately, the transition toward engineering abstraction was accompanied by a flight from practicality. As a result, a sizable fraction of today's engineering research deals with problems that are there for the sake of abstraction, difficulty, or elegance, rather than because of a real and pressing need. A similar phenomenon seems to be taking place in economics and other fields as well.

I think that this is the time to curtail the pursuit of abstraction at the expense of practicality. On the contrary, the development of abstractions that are firmly rooted in and helpful to practical problems would serve the future of our institution and of the fields that we represent in the best possible way—almost by the definition of the terms science and technology.

Thus, in technological research, our major goal should be to become more involved with practical problems without losing—indeed, while enhancing—our gains in engineering science. This goal is, in my view, reachable. For example, we could carry out a large fraction of our current research efforts in computer science in the context of a project such as the cleaning of Boston Harbor or the creation of an ideal car for the 21st century, without losing abstract vigor and while gaining stimulus and purpose from the usefulness of these projects. Involvement with such projects would also increase the teamwork skills that will be needed for success in future high-speed, high-quality-oriented industrial environments.

Turning to a broader topic, I would suggest that the Institute can have an important impact on research devoted to complex sociotechnical problems. We could start by institutionalizing commissions like the MIT Productivity Commission. There is no shortage of important issues in which interdisciplinary synergism could propel us well beyond what individual departments and schools can do by themselves. Like practical projects, such commissions would have a unifying and teamwork-enhancing effect, uniting students and faculty who might otherwise never meet.

I dream of an MIT series of sociotechnical studies on the model of *Made in America*—on the environment, on nuclear power, on biotechnology, on hunger, and on other issues—that show the world what we collectively think about all of the world's problems in which science and technology interact with individuals, nature, and society.

The experience of *Made in America* suggests that the world is hungry for and appreciative of solid knowledge on these pressing problems. It also suggests that in certain instances we should strive to communicate the results of our studies, our observations, and our opinions beyond our peers, to the public at large, especially when we collectively agree that important changes should be made. MIT should give serious thought to establishing popular lecture series on key sociotechnical issues, as well as using the broader medium of television more effectively.

Admittedly, such activities would mark a change from a tradition of analytical detachment toward a role that involves the university more actively with society. Such a shift is in my view necessary and consistent with the strategy of leading through science and technology. A greater interaction with the public would also help capture the imagination of young people and increase our ability to attract the best and brightest students to the worlds of science and technology.

Another way we could lead in tackling complex sociotechnical problems is through the creation of new research centers. The Productivity Commission has recommended the creation of an interdepartmental productivity research center that would explore issues ranging from the effect of office computers to that of new factory processes. We should also consider other sociotechnical research centers—for example, one that would make the environment its business.

Who Should Change?

Not all of MIT needs to change. I suspect that the schools that will undergo the biggest changes are engineering, management, and architecture, because they involve the disciplines and human resources associated with the "constructive arts," and this is where the greatest changes are needed.

The School of Humanities is probably next in order, because of the increasing role that that school could play, under the new strategy, in helping formulate the tech-based alternative to a liberal arts education and in helping MIT come to grips with sociotechnical problems.

The School of Science is, at first glance, least likely to be affected by the proposed strategy, since it is by idealistic definition in pursuit of pure knowledge, as in mathematics. It will be affected indirectly, however, since it is often the generator of knowledge that leads to new technologies. And it will be affected much more directly in biology, whose results are beginning to have a direct and profound impact upon individuals and the world.

A largely unexploited strength of MIT is the potential synergism among its different units. It would be a mistake for us to pursue a new strategy only from the point of view of individual departments or individual schools. Take, for example, the project of designing and building a new type of automobile for the 21st century: one that can be tailored to individual needs and is assembled in 15 minutes by snapping subsystems together, using few people, light production equipment, new materials, and new production systems. Nearly all of our departments would have to work together to achieve this goal, resulting in a heightened synergy among our peers and our disciplines. We could even drive our prototypes to and from work, giving a fresh meaning to research on reliable systems.

In short, I suspect that building intellectual bridges and teamwork across departmental and school boundaries will turn out to be as important to the future of MIT as teamwork is to the future of industry.

National or International? Government or Industry?

Universities like MIT are faced today with a host of complex questions: Should they shy away from research funded by the Department of Defense? What should they do if these funds shrink anyway? In light of congressional criticism, should they decrease their interactions with other countries? Where will the resources come from for all the big changes that would accompany a strategic shift such as the one discussed here?

Absent a crisp strategy, these and related questions are difficult to answer. Under the strategy of leadership through science and technology, the answers are a bit easier, though by no means automatic.

To lead through science and technology, we must begin by considering the ways in which we are linked to the world. The world on which our science and engineering have their impact extends well beyond the boundaries of this nation to the industries, academic institutions, and governments of Europe, the Far East, and the rest of the world—all of which are becoming ever more interdependent. In this evolving international arena, I would argue that MIT must become an increasingly international institution while retaining a primary allegiance to the United States. This means that, in peacetime, the students, industries, and educational institutions of foreign nations should be appropriate candidates for involvement with MIT. This should generally be done in accordance with the rules and regulations of the Department of State, but we must realize that these rules are not always crisp. There is an gray area in which well-meaning government agencies impose badly thought out restrictions on universities—for example, on the admission and research of foreign nationals and on the dissemination of results that are deemed sensitive for either national security or competitiveness reasons. I characterize these efforts as gray because they do not follow established laws, such as the use of the visa system, to deny entry of foreign nationals, but rather try to achieve their goals indirectly by generating a shadowy and often uninformed fear of possible leakage of technical results. Such tactics are undesirable because they push American universities into the unfamiliar territory of indirectly shaping national policy, while fragmenting university communities into tiers with distinct privileges. In this thorny area, we have an obligation to learn and educate others, including government, on what are the true issues and consequences associated with the discovery and dissemination of new knowledge.

Recent congressional hearings on the links between American universities and Japan offer a good example of the work needed to convey accurate information to the government and the public on these matters. It is widely believed, for example, that America's competitiveness problems in the 1980s were caused by other nations "stealing our inventions." But this is not true: It was not a leakage of inventions that caused the United States to lose so many competitiveness battles, but rather the inability of most of our companies to take inventions, wherever originated, and translate them rapidly and at low cost into quality products.

Restricting the flow of ideas will not increase our nation's ability to compete. And perhaps even more important, restricting the free dissemination of basic scientific and technological information will undermine our fundamental national ideology of freedom. It is, in my opinion, a fundamental misunderstanding of the nature of technology and its links to industrial competitiveness that causes people who are legitimately trying to help the United States to institute measures that actually hurt the country.

Leadership through science and technology requires that we work with both government and industry in a balanced way. At present, this means increasing the proportion of our industrially sponsored research. However, as anyone who has tried knows well, industry is relatively unwilling to sponsor research in the absence of practical short-term results. There is no question that industry must invest in longer-term research if it wishes to compete successfully in the world arena, but there is also little doubt that universities must be able to offer some tangible results if they hope to attract stable long-term funding. Fortunately, the strategy of leading through science and technology calls for a greater university participation in practical world problems and, therefore, bodes well for the production of tangible results.

There is yet another reason for becoming involved with more practical problems. Conventional wisdom asserts that the quest for practical results hampers a university's ability to do basic research. While this may be true for scientists, I suspect that it is the other way around for technologists: Playing with practical problems is often a good road to unexpected technological discovery—in the same sense that there is often more gold hidden in the mud than in pure oxygen.

Under the new strategy, our relationships with the Department of Defense and with other government agencies must go beyond the assertion of a right to funding. If we are to lead through science and technology, we must help our government to lead through science and technology. And government does need a great deal of help in this area. Take, for example, the problem of U.S. productivity and competitiveness. We might undertake to help the government exploit science and technology toward greater productivity growth and competitiveness. Our research might not change that much under such an orientation, but we would be helping in a worthy cause.

Beyond such direct help, we should also undertake to explain to the government and to the public how an internationally oriented MIT is, in the long run, far better than an insular one, not only for the world but also for the United States. As the world shrinks and countries become more interdependent, we must avoid the perils of technological isolation, and we must increasingly focus our attention on learning from the ever active global fountain of scientific and technological experience.

Thus, in dealing with the public and private sectors, our new strategy would be translated into stronger, more effective alliances with government and industry, nationally and internationally, in a quest for new knowledge in the context of more practical world problems.

Conclusion

MIT, under a strategy of leading through science and technology, would distinguish itself by providing first-class technological foundations to professionals who are not scientists or technologists; by broadening and deepening its scientists and technologists; by continuing to carry out specialized technical research; by pursuing problems of practical interest; and by tackling and speaking out on the world's complex sociotechnical problems.

MIT would redefine through its instruction and research what is meant by a professional technologist and a tech-based nonprofessional, reducing these artificial distinctions. And it would become a unique intellectual resource on those world problems that involve science and technology.

MIT would accomplish this transition not through dilution or a simple juxtaposition of current disciplines, but rather through preservation and enhancement of its proven leadership in science and technology; through careful integration and interaction among its scientific, technological, managerial, humanistic, and other capabilities; through a broader, more basic core curriculum; and through a lengthening of the time it takes to educate technologists.

This, in turn, would preserve MIT's rationale, in a manner surprisingly close to the words of its founding charter: Striving to discover and match new knowledge to human welfare through a deep understanding of science and technology, of other essential disciplines, and of the scientific and technologically based needs, challenges, and problems of tomorrow's world.

There is undoubtedly risk in undertaking this kind of a transition. Business as usual, in light of a changing world, may be an even riskier alternative.

Note

1. Michael L. Dertouzos, Richard K. Lester, Robert M. Solow, and the MIT Commission on Industrial Productivity, *Made In America: Regaining the Productive Edge* (Cambridge: The MIT Press, 1989).

The University as Quality
Alfred R. Doig, Jr.

The academic regalia in evidence at the inauguration of President Vest reminds us that the roots of our concept of higher education are close to 800 years old. We at MIT view the science-based research university as a top limb on the evolutionary tree of this educational tradition. I want to suggest that in this inaugural year it is appropriate to rethink the mission of the Institute in the context of the future of this tradition.

MIT has grown, as President Vest remarked in his inaugural address, to "become a wellspring of scientific and technological knowledge and practice." The driving force for this growth has been a relatively unlimited supply of financial resources and challenging problems. The philosophical basis for its mission was characterized by President Vest when he quoted Jacob Bronowski: "The end of science is to discover what is true about the world." Within this context, the explicit mission of MIT is "The University as Truth."

A modern science-based institution like MIT is far from being a cloistered community of scholars and students. MIT requires a complex infrastructure in order to function. To see this, consider the fact that there are close to 1,000 faculty in a total employee population of approximately 8,000; indeed, the student-to-employee ratio is itself nearly 1:1.

Over the past four decades, little attention has been directed toward the Institute's organizational structure. Current management and organizational models are hierarchical and, for the most part, simply extensions of those in place in 1950. The interdisciplinary laboratory and center concept of organization, for which MIT is noted, was a product of a wartime need. MIT has served the nation and the world well during the postwar period, and there has been little incentive to examine mission-driven organizational issues. It may now be time to consider whether the mission of MIT should be redefined to better reflect its complex nature.

In addition to internal complexities, there are many external challenges that suggest the usefulness of a mission review at this time. As the 20th century draws to a close, there is an increasing sense among those who look to the future that the world is changing in unprecedented ways. They argue that these changes will make many of our commonly accepted ideas about organizations obsolete. Consider, for example, *The Age of Unreason* by Charles Handy.[1] Handy writes that "we can no longer assume that what worked well once will work well again, when most assumptions can legitimately be challenged." He introduces the notion of "discontinuous change"—change that is nonlinear and chaotic. He argues that our current attitudes and assumptions about work, the workplace, and organizational structure are no longer adequate to cope with the challenges brought about by the dynamic nature of global change.

Handy offers a simple equation to describe organizations that respond successfully to the challenges resulting from discontinuous change: $I^3 = AV$, where I is Intelligence, Information, and Ideas, and AV is Added Value. Organizations must, according to Handy, "look ... to some of the places where knowledge has always been key and brains more important that brawn. Increasingly ... corporations will come to resemble universities and colleges. ... Universities or colleges ... are places [that] use these three Is, in theory at least, to pursue truth in an atmosphere of learning. The new organization, making added value out of knowledge, needs also to be obsessed with the pursuit of truth or, in business language, of quality."

Academic leaders might initially shutter at Handy's suggestion of a university model for business. His statement, however, derives from a conviction that colleges and universities have well-understood missions that are consistent with their environment. That shared mission has been partially responsible for the type of organizational structure that has evolved at places like MIT. It might be useful to speculate on what affect a new MIT mission statement, one I would characterize as "The University as Quality," might have on the organizational structure of the Institute.

Consider the impact of the new mission statement on Handy's equation. MIT has a higher percentage of gifted and well-trained employees than regional demographics might predict. Its working environment and reputation are largely responsible for this excellent work force. The benefit to MIT is that its employees increase the values of both the Intelligence and Ideas variables in Handy's equation. Since Handy's equation is exponential, maximizing the number of ideas and the intelligence of its employees has a powerful impact on the product, AV. In the more narrow mission of The University as Truth, only the ideas and intelligence of a fraction of the community are explicitly encouraged. Defining the mission of the university in a broader context would engage more of the intellectual capability of the whole Institute community.

A mission statement for MIT that is explicit about quality would send a motivational message to the community at large that all forms of creativity that improve the environment of MIT are encouraged and valued. This is significant because all members of an organization need to identify with the purpose or mission of the organization. "People want and need to understand the organization's purpose. They have questions, spoken and unspoken, about the purpose of the things their organization is doing, how their particular unit fits in, and what they themselves are expected to do. The organization that wants to capitalize on employees' potential for commitment therefore has to explain what it is trying to do—why it does what it does and why it is going to do things differently.... When individuals make a commitment to a corporate purpose and understand how they can contribute to it, they can call upon power beyond anything they may have employed before as they focus their energies, talents and learning ability."[2]

Now, consider the implications a shift in mission might have on the tangible and intangible outputs of MIT. The three most obvious outputs are graduates, works of intellect of all forms, and benefactors. I would argue that a mission focused on truth will have a less broadly based impact on these outputs than a mission based on quality. This is due in part to the implicit versus explicit nature of the old and new mission statements. Quality, that is, is implicit in the University as Truth, although somewhat obscured.

What might be the effect of the explicit inclusion of quality as a focus in the Institute's mission statement. For graduates, the educational experience at MIT would be broadened significantly if all employees felt connected to the Institute's intellectual environment. Employees might feel empowered to find ways of contributing to the total experience of being at MIT. The forms these contributions might take are difficult to predict since the responses involve the freedom for creative expression explicit in the new mission statement.

All employees should feel charged by the mission statement to express their ideas and contribute where they can. Graduates would leave the Institute with an understanding of total quality that would serve as a benchmark for life. Employees would have more opportunities for self-expression and self-actualization. All these contributions would add to the richness of the MIT community and the MIT educational experience.

As for works of intellect, the shift in mission would encourage a greater diversity of thought. Faculty and students would not be the only groups participating in intellectual activities at the Institute. A feeling of explicit connectedness to the broader mission statement would open new ways for members of the community to contribute ideas. After all, maximizing ideas is key to increasing added value. Employees would be encouraged to develop new methods and definitions of work, new tasks and new approaches to teamwork.

Evidence that workers are becoming more involved in redefining organizational structure is starting to appear in the press. A recent article in *The New York Times* by David Nadler describes how many companies are experimenting with a new work architecture. Nadler describes this as "self-managing units that encompass an entire end-to-end work process related to a product or service. These units are microcosms of a full enterprise, but with autonomy to structure and manage themselves. Teams work without supervisors, design their own work methods, train people in a variety of skills. . . . This has tapped into something that joins the best of teamwork and of individualism."[3]

A discussion of quality in the context of organizations coping with discontinuous change would be incomplete without a reference to the work of William E. Deming. Handy paraphrases Deming on the subject as follows: "Quality is only achieved if everyone believes in it, if everyone contributes to it and if everyone is always concerned first of all to improve their own quality at work. You get quality from quality people trusted to work positively for the good of the whole community."

How would an explicit emphasis on quality in the Institute's mission statement affect MIT's benefactors—those individuals, corporations, foundations, and other societal institutions that identify with the institution and provide resources? The impact of quality can be understood in terms of the successful salesperson's credo: A happy customer is a repeat customer. MIT is facing a future defined by finite resources and increased competition. No longer is the number of institutions with strong ties to industry small.

The change in mission also bears on MIT's relationship to its broadest benefactor: society. Gone are the days when citizens instinctively viewed the fruits of science and technology as positive. An explicit focus on quality in the mission statement of the Institute could serve to rebuild trust and reverse an increasingly adversarial relationship. The current confrontations are wasteful for all sides. It is incumbent on us to find paths out of this situation. This requires that all at the Institute find ways to build bridges to the outside, to strengthen the relationships with our benefactors. We need new and bold ideas to meet these challenges.

By making quality an explicit part of its mission, the Institute would permit broader participation and increase the probability that innovative solutions will evolve. The challenges of discontinuous change facing the Institute can be met more effectively and creatively by engaging the entire community in the process of finding new answers.

Notes

1. Charles Handy, *The Age of Unreason* (Boston: Harvard Business School Press, 1989).

2. Perry Pascarella and Mark Frohman, *The Purpose-Driven Organization* (San Francisco: Jossey-Bass, 1989).

3. David Nadler, "America, Playing to Its Own Strengths," *The New York Times*, June 2, 1991.

Toward International Education at MIT
Catherine V. Chvany

It is time we made the matter of international context and opportunity an integral part of an MIT education.
—Charles M. Vest, May 10, 1991

The Skolnikoff Report

In 1990, Provost John Deutch established a faculty study group to consider the changing international context of research universities and of MIT in particular. Chaired by Eugene Skolnikoff, the study group's charge was to advise the Institute administration on the general principles that should guide MIT's international activities and relationships and to suggest revisions in specific policies. The 41-page Skolnikoff report appeared on May 1, 1991, a few days before the inauguration of Charles M. Vest as President of MIT.

I want to start by abstracting from the report principles that bear directly on educational issues. The major conclusion is that MIT's responsibility to the nation is served first and foremost by maintenance of its position as a premier institution in education and research in science and technology. This *requires* that MIT be thoroughly engaged in international activities in science and technology—that it be a full participant in the world trade in ideas. MIT's responsibility to the international intellectual community derives from a dedication to the free and open exchange of ideas.

The rapid development of international science and technology "represents a return to the situation of the earlier decades of this century, when it was necessary to keep up with European developments in order to stay at the forefront of a field,"[1] and it is now necessary to keep up with developments in Asia as well.

Before moving on to specific recommendations, the report provides statistics on international students and faculty. The proportion of foreign undergraduates has grown from 5.2 percent to

8.7 percent in the past decade, not counting the large number of permanent foreign residents of the United States or children of immigrants whose first language is not English. These international undergraduates provide an essential element of diversity in the student body. Among graduate students, the ratio of international students is 33 percent, while about 30 percent of faculty members are foreign-born.

The decline of American K–12 schooling in science and math has created an opening for researchers trained abroad; these employment opportunities motivate many international students to stay rather than return to their native countries. Still, a great many international alumni do return home and eventually achieve positions of leadership in industry and government.

The report cites the MIT/Japan Program as an initiative of particular public policy importance. Headed by study group member Richard Samuels, it is the largest program of applied Japanese studies in the United States; it is designed "to give MIT students language and cultural skills and the experience of extended work in Japan, with the intention of modifying the imbalance in the flow of scientists and engineers and scientific and technological information between Japan and the United States. . . . MIT/Japan students will be part of the essential communication link necessary for truly reciprocal interaction between the two countries."[2] Some 250 students have completed the program since 1981. There are 46 MIT students in residence in Japan in the 1990–91 academic year.

Among the specific recommendations of the report are two critical educational needs related to the international scene, one internal and one external:

1) Strengthening the international dimension of the undergraduate curriculum to more effectively prepare graduates for the world they will face in the future; and 2) considering how MIT can contribute to improving science education in the primary and secondary schools.[3]

An Agenda for Internationalization

The task before us is to narrow the gap between a fashionable buzzword and its implementation. Plans for curricular internationalization, within the agenda proposed by the Skolnikoff report, must capitalize on MIT's unique strengths while working around and through constraints specific to MIT. The strengths that could drive initiatives in international education include:

1. A critical mass of technologically competent students, many of them of international origin, with a potential for leadership on a global scale.

2. An education built around a science core that equips every student with an international *lingua franca*.

3. Seven programs in Foreign Languages and Literatures—Chinese, French, German, Japanese, Russian, Spanish, and English as a Second Language—under a single umbrella rather than in traditional literature-driven departments.[4]

4. The vision of MIT's founders—*mens et manus*—that so creatively transcended the traditional opposition between education and training.

These strengths are coupled with daunting demographic and cultural constraints:

1. Outside of the MIT/Japan Program, less than a dozen undergraduates annually enroll in term-time foreign study. As far as I could ascertain, only a few more can afford to give up summer earnings for a summer abroad or in stateside language immersion programs. A crucial element in the success of the Japan Program is clearly the prospect of paid employment in Japan.

2. While a junior year abroad is the norm at several institutions with which MIT competes for versatile students, MIT's four-year engineering programs hardly permit interruption for a term of foreign study.[5]

3. The Foreign Languages and Literatures offerings, however suitable their skeleton, are badly in need of fleshing out.

4. Strange as it seems in the home of *mens et manus*, the usefulness of foreign languages has raised perennial doubts about their humanistic value.

As often happens at MIT, programs and opportunities exist *de facto* as individual initiatives, long before they exist *de jure*. For years, MIT undergraduates have managed—within and around the four-year curriculum, or in combination with the first year or two of graduate study—to prepare for leadership in international science, technology, or public service.

MIT has always hosted a few students who enter with a head start, who can afford summer study or study abroad, and who are energetic enough and sufficiently motivated to excel in science or engineering as well as in foreign languages or area studies. A few complete a second or even a third major while doing graduate work in their primary field. A few graduate students have persuaded their advisors to let them minor in a foreign language. The number of students engaged in such programs could probably be increased dramatically simply by improved publicity inside and outside MIT, but the percentages would still be too small to be typical of "an MIT education."

MIT must find ways of offering international education to a larger group of undergraduates, and opportunities must extend into the graduate years as well. The level of competence necessary for leadership in international science and industry requires that excellence in science or engineering be coupled with the equivalent of a first-rate major in a language area. Even for MIT's overachievers, this requires extra time. Specially designed graduate fellowship programs are needed to buy time for students to meet these dual objectives—and to motivate schools and colleges to groom suitable candidates.

First Steps and Next Steps

The utilitarian thrust of the MIT/Japan Program is coupled with ongoing efforts to integrate Japanese into MIT's humanities offerings. A minor in East Asian Studies has been proposed. Like the older Russian Studies and Latin American Studies, this low-cost area program is made up of existing courses from various departments, to be capped eventually by a seminar. From 1991–92 on, the Japanese language subjects will be complemented by a subject in Japanese culture and literature taught in English (as in other such courses, qualified students will be encouraged to do some of the work in the original language).

It may surprise some that MIT had majors in Japanese and East Asian Studies long before MIT offered Japanese, through the device known as the Major Departure (within the "joint" majors in Humanities and Engineering or Humanities and Science or the "full" major in Humanities). For years students have cross-registered for Japanese and Chinese at Harvard or Wellesley, have found suitable subjects in other MIT departments, or have transferred credits from elsewhere. Three early majors are listed in the *Course XXI Survey and Register 1958–1983*, and there have been several more since. Most of them also had a full designated major in a science or engineering field.

In the modest words of the Skolnikoff report, the MIT/Japan Program "is a small start given the size of the national need, but appreciable for a single university."[6] The Japanese numbers are spectacular, however, when compared with MIT's statistics for foreign study in Europe. While some students have gone on to international careers in European areas, preparation has been left to individual initiative.

In the aggregate, enrollments in foreign languages are healthy—second only to economics. Extracurricular support is provided in associated living groups—French, German, Russian, and Spanish Houses in the West Campus New House, each with a graduate resident tutor fluent in the language. While French, German and Spanish offer fewer job incentives than Japanese, they do have sizable populations entering at intermediate levels or above. High-school French and Spanish are still strong in the United States, and the reputation of German as an important language for scientists brings to MIT many freshmen who can enter second-year German. Freshmen entering with these languages often continue them at MIT for at least a term.

In MIT's packed curriculum, foreign language enrollments are immediately affected by changes in requirements, especially the eligibility of language subjects for distribution credit and for the new minor. Growth during the decade 1974–84 (counter to national trends at the time) became possible when second-year foreign-language subjects were finally included in the humanities core requirements. During that period, MIT's foreign language programs achieved national leadership and international prominence. Enrollments plummeted after 1985 (again counter to national trends) as an unexpected consequence of curricular reforms, which withdrew the distribution option in third-term language.[7] The severity of the losses in West European languages is partly masked by the addition of Japanese and now Chinese. By fall 1989, second-year language subjects were again granted distribution status, but new freshman rules introduced a new handicap to language study: The reduction of maximum freshman course loads essentially replaces the option of a 12-unit language with a 6-unit seminar in some other subject.

The new minor has, so far, attracted 19 in French, 20 in German, 14 in Spanish, and 8 in Russian. The small numbers reflect a handicap unique to languages at MIT: While all other humanities subjects feed into 6-subject minors, minors in foreign languages begun at MIT require eight subjects.[8] For freshmen entering language at a second year-level, the fact that the first two terms do not count for the minor is not a serious deterrent; but for languages not taught in high schools, or for any student starting a new language, access to the foreign language minor (and thence to the major or to most foreign study programs) depends entirely on whether first-year language subjects can be crammed in during freshman year.

Since majors in straight language-and-literature areas average one per language per year, MIT should obviously not attempt to provide comprehensive majors covering all centuries of French, Spanish, German, or Russian literature. Area studies seem more feasible and cost-effective. There are enough offerings throughout MIT for a Russian Studies program; but there are currently not enough resources to assemble separate programs in, say, French Studies or Germanic Studies. The section is now considering such possibilities as European Studies. Latin American Studies, which had languished since the departure of the political scientist who started it several years ago, now incorporates subjects in peninsular Spanish literature and the history of the Spanish language (cross-listed in Linguistics).

Russian Studies has drawn an average of two majors a year, in addition to one major in straight Russian. The relative success of Russian Studies/Russian as a major is surprising in view of Russian's multiple handicaps. Freshmen entering MIT with Russian have gone from 60 in 1967 to at most a half-dozen now. Russian is more difficult for Americans to learn than are French, German, or Spanish. The almost entirely home-grown pool of advanced students is therefore small, limiting post-second-year offerings in the language. Russian makes up for these handicaps, however with support from rich offerings in other units, notably Loren Graham's courses in the history of Soviet science, and—until recently—offerings in the Literature section.[9] Present plans include conflating the present Russian minor and Course XXI majors into a single more coherent Russian Studies program. The incoming Coordinator of Russian Studies, historian Elizabeth Wood, will design a capping seminar with participation from interested faculty throughout the institute.

While only a small percentage of students have so far managed to combine a science or technical major with a second major in a language or area studies program, the majors in Russian Studies or Russian comprise a high percentage of the very best students, those elected to Phi Beta Kappa, those who win prizes and the most coveted fellowships to graduate schools.

Even before the establishment of Russian Studies in 1974, students incorporated Russian area studies into existing programs, or designed a Russian-related major as a Major Departure. A few more took enough Russian-related subjects for a double (or even triple) major by present rules, even though they did not declare it. Several are now internationally known scholars, enriching fields in which scientific literacy is a rare and precious commodity, such as Slavic Literature, Linguistics, and Russian History.

An early student who could well have declared Russian as a second major was Edward F. Crawley '76, one of the first handful of undergraduates to participate in the Leningrad University exchange sponsored by the Committee on International Educational Exchanges (CIEE), a consortium of liberal arts institutions. Early pioneers like Crawley were admitted to the CIEE exchange in nationwide competition with Russian majors from liberal arts colleges. Now a tenured member of the Course XVI (Aeronautics and Astronautics) faculty, Crawley has been instrumental in arranging an exchange with the Moscow Aviation Institute that requires less language preparation, for the MAI is more interested in the students' engineering than in the elegance of their Russian.[10] He has agreed to share his knowledge of Soviet engineering and aviation with the future Russian Studies seminar.

Several patterns emerge from student case histories:[11]

"Joint" XXI Major as Second Major or Only Major

The majority of those who take a joint major (XXI-E or XXI-S) do so in addition to a full designated major in a science or engineering field. A few take only the joint major. The flexibility of the latter option seems to appeal to students who go on to prepare for careers in religious ministries, law, or medicine.

The Five-Year (or More) Mode

Several MIT graduates completed their second (or even third) major in XXI—while working on advanced degrees here or elsewhere. Their experience can serve as the beachhead for a five-year program of engineering for international leadership. MIT might consider extending this option to more graduates, and even to non-MIT graduates with strong undergraduate humanities.

Full Major (and Minor) Departures

In several humanities fields, the full XXI major is only possible with help from transfer credits from foreign study and/or from cross-registration with Harvard or Wellesley.[12] The flexible Major Departure has allowed students to choose a (second) major in a field not offered at MIT. From the beginning of Course XXI in the 1950s, a few students have designed full or joint majors in Hebrew, Latin, Japanese, Russian, Scandinavian, as well as concentrations or minors in fields not taught at MIT.[13]

Self-Designed Breadth

Most Russian Studies majors chose thesis topics in which they could combine their scientific or engineering training with their interests in Russian language and culture, with topics such as: the rhetoric of scientific articles across cultures; Soviet press coverage of environmental issues; coverage of computers in Soviet popular science magazines; the language of the ABM treaty; topics in the history of Soviet science; and several translation projects, including science fiction.

The Double-Duty "Physics" Thesis

The Physics department requires a thesis for the S.B., but it does not have to be in physics. For some students with a double major in physics and XXI-S, the thesis does double duty. One or two students with double majors in, say, physics and mathematics, who did the "physics" thesis in a language field, could have qualified for a third S.B. in XXI-S if they had declared it.

The Graduate Minor in Foreign Language or Area Studies

Here again, student initiatives have been ahead of established programs. MIT could affirm a commitment to internationalization simply by approving and announcing such an option.

Preparation for Graduate Programs Requiring Two Foreign Languages

Several students majoring in one language area take a minor in a second language. Many other students planning graduate work in international business or in social sciences need to acquire not only literacy but competence in two foreign languages and the cultures in which they are spoken.

Needs and Resources

The Skolnikoff report's final recommendations reflect a recognition that internationalization cannot be separated from the precollege agenda. Given the thin curricula of U.S. high schools, where juniors and seniors typically have to choose between advanced levels of science and foreign languages, MIT's natural bias in favor of science militates against selection of students with strong foreign language. Whatever messages MIT sends to the primary and secondary schools, it must take special care not to pit science against foreign languages or area studies. International science and access to international cultures are both necessary. Signals that languages as well as science are valued by college admissions officers can provide much-needed support for precollege language offerings.[14]

Given MIT's curriculum, nearly all the students who have been able to develop an international dimension are those who came to MIT with a head start, in core requirements, in foreign languages, or both. By expressing a strong preference for such versatility, MIT could provide support for schools that offer strong preparation in both science and languages.

I now return to MIT's strengths and constraints. Two of the pairs I have noted relate to MIT's demographics and its curriculum. The last two concern the marginal status of foreign languages at MIT, and institutional cultures that are at best ambivalent about internationalization.

The first pair reflects the fact that MIT is traditionally not a school for the rich. MIT's needs-blind admissions policy continues to provide many with a ladder out of relative poverty. But then the development of the same students into international leaders cannot be left to family resources. If MIT is to offer an international education to those outstanding students it attracts from disadvantaged school districts, it must loosen up restrictions on language study by freshmen. For example, MIT might waive the freshman unit limit for students enrolling in language subjects, *in conjunction with a waiver of expectations for term-time earnings*. Summer study in the United States (including transfer credits from less expensive state institutions) and abroad (with transfer of financial aid) must be encouraged with waivers of summer earnings as well as with scholarships and work-study opportunities like those of the MIT/Japan program or the Moscow Aviation Institute.

MIT's large population of international students is a largely untapped resource which could be better integrated into the education of undergraduates and graduates. The great majority of students in MIT's program in English as a Second Language are graduate students.[15] Many of them will eventually hold positions of power in their native countries, where they will be considered to be experts on the United States. This population should be encouraged to develop competence in both American English and American Studies with the option of a graduate minor. Such a graduate minor should not, however, be limited to international students, for Americans could join such a program for mutual enrichment.

The second paired strength-and-constraint involves the four-year undergraduate curriculum and the culture that supports it. A number of undergraduates have managed to design their own versions of five-year programs leading to double or triple degrees. For some the "fifth year" came before MIT, for others it extended into graduate school.[16] MIT must open up opportunities at both ends. In addition to supporting strong precollege programs and attracting more students who can enter with a year's head start—many of whom now choose to go elsewhere, MIT must also offer flexible opportunities for continuing international studies concurrently with graduate work.

The more general issues will probably be more difficult to address. While the common umbrella of Foreign Languages and Literatures has certain advantages in the abstract, the concrete situation at MIT places the language fields at a disadvantage. In any liberal arts university of comparable quality, the humanistic value of languages would not need defending. Each language department would have allies in neighboring departments with common interests and an understanding of differences. The languages we teach are inherently different, they present different difficulties for Americans, their cultures differ in their literary achievements, in their past or present political and economic importance, and in the scholarly traditions that have developed within them. MIT must affirm the humanistic as well as the practical value of foreign language study and facilitate (rather than handicap) the election of languages by students at all levels; it must also avoid the temptation of across-the-board solutions that force different language programs into lockstep.

The major block to internationalization is that excelling in science or engineering as well as in language-area studies adds up to two full-time jobs. While students do manage such loads in surprising numbers, MIT's culture favors single-minded devotion to one field. Graduate students who take language subjects are often advised to audit without registering, so that this deviant behavior will not appear on their records. Professional science is too demanding, too competitive, for such a major distraction, unless it is legitimized by a fellowship program that does not eat into the science department's shrinking funds for research and training.

What is needed is a cooperative funding venture between humanistic education programs and those agencies which support training in science and engineering, whereby the first provide one-fourth of a graduate student's support for four years, buying the student an additional full year of graduate study. This should be a frankly elitist program, geared to candidates identified as "future leaders in science, industry, or government," who have demonstrated interest and aptitude in foreign language-area studies during their undergraduate years. Such students could achieve true competence in a foreign language area while excelling in their doctoral work in a science or engineering field. The existence of such a program would stimulate talented undergraduates and even high school students to begin language study as early as possible in order to qualify for it.

Ideally, the "future international leaders" program would involve 10 American first year graduate students annually, while a parallel program in American Studies would be available for five future leaders from among the international graduate students. Both programs might be capped by a joint seminar focused on the governments, customs, and laws of the respective countries.

Moving beyond token opportunities will require efforts on several fronts at considerable expense: major fellowships, waivers of term-time earnings, Tech Abroad programs, more courses and faculty to teach them. That is the easy part. The more difficult task is to transcend the isolationist tendencies in the national culture, and an institutional culture that perceives students who move on to humanistic studies as lost. One hopes that, as MIT considers new educational initiatives, including recruitment, it will seek the counsel of those alumni and alumnae who successfully designed educations that integrate science or technology with international studies. Somehow MIT must learn to see these losses as gains, not only for the humanistic fields, but for international science and technology.

Notes

1. Eugene Skolnikoff et al., 1991. *The International Relationships of MIT in a Technologically Competitive World*, report of a Faculty Study Group appointed by the MIT Provost, 6, n. 5.

2. Ibid., 38.

3. Ibid., 40.

4. The 10 members of the Foreign Languages and Literatures faculty also participate in small interdisciplinary programs such as Cultural Studies, East Asian Studies, Film and Media Studies, Latin American Studies, Russian Studies, Studies in Language, and Women's Studies. A pilot program in Mandarin Chinese will be launched in 1991–92 as part of East Asian Studies, with the help of a three-year grant from the Chiang-Ching-Kho Foundation.

5. Opening up foreign study opportunities would require special Tech Abroad programs supervised by MIT or by a consortium of engineering schools. Almost no existing study-abroad programs offer engineering in the foreign language, and the few that offer some science rarely match the quality and rigor of MIT's offerings.

6. Skolnikoff, *International Relationships*, 38.

7. Between 1985 and 1989, MIT's language enrollments dropped from a high of 1,896 to 1,497. In view of national trends showing a 50 percent *increase* in foreign language enrollments for the same period, the loss of language students at MIT is far greater than the actual drop of over 20 percent. Some of the antilanguage effects have now been corrected and enrollments are rising slowly, but they are still far below 1985 levels.

8. Inexplicably, while six courses in history or literature (without any foreign language) are deemed to provide competence worthy of a minor, six courses in a foreign language are not—even though successful completion of three years of foreign language enables a student to survive in an academic program abroad.

9. Some of the losses in Russian Literature are due to the death of Professor Krystyna Pomorska in 1986 and the retirement of Professor Robert E. MacMaster in 1990.

10. The summer exchange with the Moscow Aviation Institute was launched in 1989 by Crawley and other members of the Course XVI (Aero-Astro) faculty. Starting in 1991–92 the summer exchange will be supplemented by a term-time exchange, but since the U.S. partner must pay full MIT tuition to cover a Soviet counterpart at MIT, even the modest quota of four may be difficult to fill. The MAI has many more candidates hoping to come to MIT.

11. This based on Travis R. Merritt, ed. 1984. *Course XXI Survey and Register 1958–1983* (Cambridge: MIT Humanities Undergraduate Office), and on informal student records since then.

12. In 1975, when the old Course XXIII was split into Foreign Languages and Literatures (which went with Humanities) and Linguistics (which joined Philosophy in XXIV), MIT even gave serious consideration to doing away with foreign languages altogether and relying entirely on Wellesley and Harvard cross-registrations. In practice, MIT student schedules preclude regular atten-

dance of classes at distant campuses. Wellesley students find it easier to attend classes at MIT than vice versa.

13. For such students, MIT offers a Studies in Language concentration that includes two languages at fourth-semester level or higher plus a Studies in Language or Linguistics subject. The possibility of a minor and XXI major in cooperation with Linguistics is currently being explored. Wellesley College offers an interdisciplinary Language Studies major whose students take MIT Studies in Language subjects.

14. By the same token, the wholesale dropping of college language requirements in the 1960s gave the high schools license to follow suit. As a result, high-school Russian all but died, removing the market for the College Board exam, which in turn made it harder for the surviving programs to justify themselves.

15. English as a Second Language was begun 25 years ago as a minimal service program under the sponsorship of the Medical Department, out of concern for the stress many foreign students suffered when they came here to study with minimal abilities in conversational English. Admission standards for international students' English have since risen.

16. According to the *Course XXI Survey and Register*, 64 percent of the 528 humanities majors responding in 1983 had earned at least one advanced degree. Of the 118 Ph.D.s responding (from a 1958–83 pool of 847), two-thirds were working in their humanities field, and one-third in their science or engineering field.

The Research Library in the Information Age
Jay K. Lucker

There has been a great deal of speculation, both written and oral, about what the academic library of the future will look like. Visions of libraries in the 21th century include one that foresees the entire universe of recorded knowledge available in electronic form thus eliminating any need for physical storage of information. At the other end of the spectrum is the idea that libraries will continue to acquire large amounts of printed information and to organize and preserve much in the same manner as presently occurs. These are obviously both extremist and impractical views of the evolution of libraries at institutions of higher education, but both of them contain elements that I strongly believe will characterize the research library in 2000 and beyond.

In this article I intend to do the following: identify the key factors affecting the way that research libraries support collections and services today; cite some of the changes that are taking place and the key factors causing these changes; and finally, outline the principal components of a vision of the MIT Libraries as a prototypic research library in the 21st century. My thesis is that predictions of the imminent demise of the university library are misguided and quite premature. Libraries have been one of the most enduring components of the academy, historically occupying a symbolic role as the "heart of the university." On many campuses, the main university library is located at the physical center as well. Speculation regarding the survival of libraries has been generated by the increased availability of information in electronic form as well as the proliferation of personal computers, mainframes, and local, national, and international networks. A view exists that all knowledge worth storing can be accommodated in large databanks and, thus, there will shortly be no need for repositories of printed and graphic information such as presently exist in academic, public, and specialized research libraries. There is no question that on-line access to information has had a revolutionary effect on libraries as well as on their users. There has been tremendous change in libraries during the past two decades and there will continue to be significant changes in the concept of what defines a library's collection, of how and where information and reference services are delivered, and of how to measure the effectiveness of an academic library in support of education and research. What will undoubtedly remain as part of the library of the 21st century is also significant: the library as a place for learning, reflection, and creativity; the vital importance of the librarian as an information expert; and the requirement for the library to respond dynamically to information needs of students, faculty, and research staff.

Library Collections and Services

The historic role of the research library has been to collect, organize, and preserve recorded knowledge as it supports the information requirements of current and future students, faculty, and researchers. Libraries have been assessed primarily upon the basis of how much material they have actually amassed and stature based on the idea that "bigger is better." Many libraries have set as a goal the development of comprehensive collections in specific fields—comprehensive meaning all materials in all languages—with the idea of being able to respond locally to whatever users required. Not only were current needs expected to be satisfied but also future needs of then unidentified (and perhaps unborn!) generations of users. The idea of the complete collection was seldom realized. The largest research libraries like the Library of Congress and Harvard University Library certainly acquired broadly and deeply in many areas but even as early as the mid-1950s, it was evident that complete coverage of a subject or a language or a period was impossible to attain. The major obstacle to building comprehensive collections is certainly economic—library budgets are simply not large enough to support such a level of acquisition and they have been eroded steadily in purchasing power by a level of inflation that has consistently exceeded both the Consumer Price Index and the Higher Education Price Index. There has also been a steady, almost geometrically expanding rise in the amount of scholarly information published. Nonetheless, there remains with some members of the academic community and, indeed, with a few librarians as well, a residual belief in the "bigger is better" concept.

How have libraries responded to the double pressures of an increasing volume of published information and the escalating cost of acquisitions?

- Increase the library budget;
- Across the board reductions in acquisition level;
- Selective cancellations of specific titles;
- Resource sharing with other libraries;
- Greater reliance upon commercial document delivery services;
- Reallocation of funds from other parts of the library budget such as staff, services, or operating expenses.

In practice, over the past decade or so, many large academic research libraries have employed all of the above but today still find themselves buying fewer books and journals even while acquisition budgets increase.

There are a number of other pressures acting on academic research libraries:

1. There is continuing growth in the number and diversity of interdisciplinary studies like energy and environment, women's and men's studies, African-American studies, ethnic diversity, biotechnology, and real estate development. Among the characteristics of these areas is that they are heavily literature dependent, cross traditional faculty *and* library demarcations, and tend to add to rather than replace existing specializations.

2. The information technology revolution requires a huge investment in equipment, software, and space but electronic data systems supplement rather than supplant earlier technologies like print and microform. Libraries have been and continue to live in a bimodal environment.

3. Despite limitations on growth engendered by fiscal constraints and by electronic storage, libraries require additional space for expanding collections as well as for new functions and programs. There is a limited set of options for most research libraries: build a new library; add to existing facilities; renovate non-library buildings; utilize on-campus or remote storage facilities.

4. The role of librarians is changing dramatically with the advent of new technologies and the globalization of information. In addition

to a broad general education, specialized subject knowledge, and professional library expertise, librarians must be information technology literate. There is also a need for more people with analytical and managerial talent.

Ingredients for Change

The academic research library is caught up in a whole set of changes affecting both the parent universities they serve and the world of higher education in general. Among the key factors are demographic patterns, diversity and affirmative action, government support for education and research, and the application of computer and other technologies. Of these, it is technological change that has had and will continue to have the most lasting and widespread impact on the library.

The development of a national on-line bibliographic system has enabled libraries to share the burden of cataloguing through the establishment of standards and the utilization of a single bibliographic record by multiple institutions. Centrally stored records can be transferred to local systems and used to support individual library on-line catalogs. The national database serves not only as a cataloging tool but also as a union catalog for locating items and requesting interlibrary loans or photocopies.

The creation of subject-oriented on-line databases and information systems has completely changed the way that scholars and students in most fields undertake research. On-line files that abstract and index published information are commonplace. Databases that contain the full text of journals, newspapers, technical reports, patents, and laws, are becoming more numerous. Databases with text and graphics and that are capable of manipulation and computation are projected for the near future.

The personal computer has become ubiquitous. It permits individuals to create documents in electronic form. It serves as a gateway to local and remote networks. Through the campus network, the PC connects the user with a local library catalog, with campus computers and local databases, with national bibliographic databases, and with electronic mail systems, newsletters, and bulletin boards.

Since a vast majority of published information of interest to the research university community is generated in machine-readable form either by the author or the publisher, the capacity exists for the publication of full text in electronic form. Within the next 10 years, major changes will inevitably take place in the production and dissemination of scholarly information, especially in science, medicine, and technology where the cost of printed journals is the highest. There is currently, however, a great deal of uncertainty regarding peer review, refereeing, graphics, standards, access, copyright, charges, and archiving. I believe that there will be a slow increase in the number of electronic journals over the next 10 years but that it will be well into the 21st century before a majority of scholarly journals in all fields appear in this form.

The development of CD-ROMs (compact discs with read-only memory) for library applications has made available a large number of databases at a relatively low cost. A single CD-ROM holds some 500 megabytes of information and can store entire reference works such as encyclopedias, handbooks, statistical compendia, and directories. CD-ROMs are also being widely used for abstracting and indexing services and for full text storage of journals. The advantages of CD-ROM include relatively low acquisition, equipment, and operating costs plus a wide range of electronic searching capabilities. Drawbacks include slow access and search speed, single user access, and concern about long-term archiving of data. The development of local area networks and multidisc towers has ameliorated some of these problems.

Among the most dramatic changes taking place in higher education is the concept of the "wired campus." The linking of academic and administrative departments, faculty offices, classrooms, laboratories, dormitories, *and* libraries to a university-wide telecommunications system has been the critical factor in promoting the concept of the library without walls. Library catalogs were one of the first databases to be made available on many campus networks. These have been followed by directory services, abstracting and indexing services, and full text files accompanied by electronic mail, on-line reference librarians, and document ordering and delivery. Concomitant with these developments has been the growth of library and information services available through national and international networks. Services like BITNET, DECnet, and Internet now provide members of the academic community with access to an increasing number of catalogs of individual libraries and library systems.

The application of information technology in individual research libraries has had a tremendous impact both on the way that libraries organize themselves and their services and on the library's users. There is almost no aspect of library activity that has not been affected by computers. In addition to the use of national databases for cataloging and interlibrary loan, libraries have developed on-line acquisition systems in which locally created orders are transmitted directly to book suppliers. In these systems, order records are simultaneously loaded into local on-line systems and fund accounts are encumbered. On-line records can be updated to show the progress of the item through the system. The automation of library circulation processes has resulted in increased speed and efficiency, lower staff costs, and a great deal of helpful information about usage patterns. Local library systems are also being used to support other bibliographic and full text databases including abstracting and indexing services, research directories, and full text files of technical reports and theses. The key element in local systems is the on-line catalog that provides a number of distinct advantages over manual files. All of the data elements in a record can be searched either separately or in combination (Boolean searching). Circulation status can be displayed with individual titles and volumes. Information about material location can be changed easily; this is especially helpful in libraries with on-campus or off-campus storage facilities. The on-line catalog can be expanded to include additional information about items in a library's collection such as tables of contents, index entries, and additional keywords. Libraries can also add information for individual titles and collections to which they have guaranteed access but do not necessarily possess, such as might be available under consortia or cooperative agreements. There is great pressure upon research libraries to convert records for all of their holdings to electronic format. Since most libraries did not begin cataloging on-line until the 1970s, this presents a major challenge. The existence of the national bibliographic system is of particular importance in this regard as it provides shared access to converted older records.

While access to information about the existence of a particular book, thesis, or periodical article is indispensable for learning and research, it must be supported by an effective delivery system. The use of telefacsimile for the transmission of requests and for full text delivery has expanded dramatically during the past five years. The expansion of digital fax will provide direct delivery to individual workstations and there has been considerable progress in the development of scanning and digitizing from bound volumes.

The Mission of the Research Library

The traditional mission of the academic research library has six components: acquire, organize, store, identify and deliver, interpret, and preserve. In the MIT Libraries, we have shaped our mission to reflect the special characteristics of the university we serve:

> To provide high quality services and collections to meet the needs of MIT's education and research programs. To provide a place conducive to discovery and self-education outside the classroom and laboratory. To share with the scholarly world at large the unique information resources of the MIT Libraries. To take an active role in cooperative efforts that ensure access to and preservation of information for scholarly research.[1]

Inherent in this statement are a number of assumptions. One is that libraries are not self-sufficient. Our ability to provide access to information must be linked to national and international programs in information technology and scholarly communication. A second is that future progress depends in large measure upon our ability to apply technology effectively both to fill the information needs of users and to control the costs associated with the operation of a large research library. It is further assumed that, for the foreseeable future, libraries will continue to maintain large print collections, particularly in support of literature-dependent disciplines, while simultaneously moving ahead in providing technology-based information services.

The principal efforts undertaken by the MIT Libraries to support its mission include: (1) to serve as a gateway to national and international information resources; (2) to provide access to information and materials in any format, especially electronic formats; (3) to provide a variety of services geared both to sophisticated and unsophisticated information users; (4) to develop and manage collections of materials which support the research and educational programs of the Institute; (5) to organize, arrange, and preserve materials in logical and rational ways so that they can be found and used; (6) to maintain an administrative structure that allows efficient functioning; and (7) to provide an environment in which the Libraries' staff can flourish.

Directions for the Future

Research libraries, as they chart their future course, will have to be cognizant of and responsive to decisions and trends over which they may have little or no influence. Some of these come from the larger world of higher education and technology; other from the institution served by the library; and others, from within the library itself. For an environment like that of MIT, several major trends were identified as part of a recent strategic planning process.

1. Identify user needs. Faced with an exponential increase in the amount of published information and in a variety of forms, research libraries will have to identify and respond to the particular needs of local clientele and to structure, market, and deliver services responsive to these needs.

2. Emphasize access to information. There will necessarily be a greater emphasis on access to information and less of an attempt to build comprehensive local collections. Rapid delivery of information, increasingly in electronic form, will become a critical measure of the effectiveness of the library.

3. Exploit new technology. Newer technologies will provide faster and more comprehensive access to information but will not quickly replace existing formats and technologies. This will increase the overall cost of operations and require innovative budgeting and new sources of funds.

4. Exploit local networks. The availability of personal computers and campus networks presents both an opportunity and a challenge as faculty, students, and research staff become more information-literate and their needs become more sophisticated.

5. Link library programs to academic programs. With limited resources and an ever growing universe of information, there must be a close mapping between library collections and services and the educational and research priorities of the university.

6. Monitor economic pressures. For the immediate future, a variety of economic pressures will cause a steady erosion in the buying power of academic libraries: general inflation in the cost of scholarly information, especially serials, and in operating costs; a spiraling number of new serial titles and their introduction in new formats; fluctuations in the value of the dollar abroad; the decline of government sponsored research.

7. Reallocate resources. There will be a continuing shift of budgets away from staff and collections, toward the support of automation, telecommunications, and contractual arrangements for information access.

The MIT Libraries at the Beginning of the 21st Century

Historically, the MIT Libraries have collected and preserved recorded knowledge most relevant to the paths of inquiry taken by faculty, students, and researchers at MIT. With the volume of relevant printed information increasing, and with more and more information important to our needs being produced in electronic form, and with increased demand for quick and direct access by the end user, it is clear that new roles and relationships are required for the research library in 2000 and beyond. It is also likely that many of the characteristics of research libraries that exist today and that have endured for centuries will also be a part of that era.

The Library as Place

Library buildings will continue to serve an important role for housing physical collections of information with convenient spaces for users to consult those materials. Libraries will continue to serve as locales for self-education and discovery outside of the classroom and laboratory; they will continue to be a haven from the pressures of academic life and communal living. They will be a place of particular importance to students as a part of the social and intellectual experience of an MIT education.

Scholars will depend increasingly on the library for access to printed information. Despite the growth in electronically published research, the number of paper-based books and journals will continue to rise. In keeping with recent trends, the amount of printed information acquired by individual faculty and research staff will continue to decline. The library will also serve as the locus of information about new technologies and as an access point for specialized information resources.

The number and scope of small departmental libraries and reading rooms outside of the formal library system will continue to decline because of economic and space pressures. They will be replaced by electronic library modules consisting of small collections of current journals complemented by electronic access to local and remote bibliographic and full text databases, by on-line communication with library specialists, and by expeditious document delivery service.

The catalog will contain records for materials in the Libraries' collections including archives and manuscripts, maps, slides, machine-readable data files, and software, as well as books and journals. Through regional and national networks and cooperative preservation programs, library users will be guaranteed access to any research materials required.

Extending the Boundaries
The concept of the "library without walls" will be fully implemented. At any time of the day or night, the library's catalog and other on-line information services will be accessible through workstations in faculty offices and in student living areas. Among the databases available will abstracting and indexing services, full text files of journals, a directory of campus research linked to published technical reports and articles, and a visual images file representing material in the library's slide collection. Users will also have electronic mail access to subject specialists and to an expert system research interface. They will be able to download information to local workstations or request delivery of hard copy.

Through connections with regional and national networks, users will be able to browse catalogs of other libraries as well as bibliographic and full text databases. Information in these databases will be textual and visual, in completed form or in progress, and users' comments will be able to be incorporated into the files. There will extensive files of statistical data available that can not only be downloaded but also manipulated at the local workstation.

The application of hypermedia, simulation, expert systems, and other sophisticated technologies, will cause dramatic changes in the role of information in education and research. Research libraries will continue to expand their role in promoting the utilization of resources by students and faculty and in developing new relationships with information providers throughout the world.

The Role of the Librarian
Research librarians will be an indispensable element in dealing with the complex world of information in the 21st century. They will combine their subject knowledge with their skills in database development, information storage and retrieval, and management, to identify and deliver what library patrons need.

Librarians will work closely with research groups and with individual faculty to identify and evaluate information resources, and to keep them aware of new publications. While direct access to databases by the end user will increase, the librarian will play an essential role as expert guide through the maze of networks and files. Through course related and course integrated instruction, librarians will assist students in learning about the structure of information in their major fields, in finding cost-effective strategies for retrieving that information, and in developing information gathering skills that will serve them in later life.

The New Research Library

By the year 2000, most internal library operations will be automated and staff members will use individual computer terminals to carry out their day-to-day responsibilities. There will be close collaboration among the library, academic computing, and administrative computing in the design and delivery of information. The campus network will provide a wide-range of services, many of which will be developed in the library. There will be a international network of research libraries with each institution taking responsibility for acquiring and organizing locally created information and for providing worldwide access to unique and specialized material in its collection.

The pace of change in the 21st century will require great flexibility in the way that libraries organize themselves and utilize their staff and resources. Measures of quality of research libraries will transcend the traditional counting of volumes and titles and will instead look at their ability to match user needs with relevant information. Libraries will continue to exist as physical entities with collections, staffs, and services. They will, above all, be a vital factor in fulfilling the mission of the university in creating, preserving, and disseminating knowledge.

Note

1. This statement and other sections of this paper have been adapted from *The MIT Libraries at the Beginning of the 21st Century—A Strategic Plan* (Cambridge: The Libraries, Massachusetts Institute of Technology, 1988).

Electronic Organs of the Mind: Language and Computation for the 21st Century
Robert C. Berwick

Introduction: Subway Science

Thirty feet underground, in the Kendall Square subway station that now, symbolically enough, runs right to the very edge of MIT itself, the wall is tiled with a memorial time line celebrating technology and MIT. Set against gray and maroon ceramic, the glossy plaques commemorate the many achievements of *mens et manus*, mind and hand: LORAN, time-sharing computers, the artificial gene, magnetic core memory, information theory. Curiously enough, on the more than 100 plaques there is but fleeting mention of perhaps the most revolutionary MIT contribution of all: transformational generative grammar. The realization that our brains come into the world tailored for language—with a "language organ" much like the heart or liver, but built for all human languages—launched the modern cognitive science era and the scientific study of the true, inner connection between mind and hand. Even so, the underground time line stops with that one brief mention of this revolution some 30 years past.

Linguistic science at MIT did not stop, however. In fact, the same linguists who made the first revolution have recently made a second, once again fundamentally altering the links among cognition, language, and, this time, computation—mind, hand, and now machine. This second revolution has shown that languages as seemingly diverse as English and Japanese can be assembled out of 20 or so simple building blocks, combined in different ways just as atoms combine to form seemingly diverse chemical compounds. What this essay will offer, in lieu of a new subway plaque, is an overview of the astonishing implications of these findings for computers and language in the 21st century.

These discoveries will have a growing importance for our ever-shrinking multilingual, networked world. Already, researchers in the Artificial Intelligence Laboratory and the Linguistics Department have used these language building blocks to create an electronic version of the language organ—a computer system that, in the twinkling of an eye, can analyze Japanese instead of English without being reprogrammed.

Linguistics, Computation, and the Tower of Babel

As in any revolution, to get this far we had to discard some old, sometimes cherished intellectual baggage—such as the notion that languages are composed of many thousands of rules, surface patterns like the ones we met in (that aptly named) grammar school. In this traditional view there are distinct "rules" that spell out in detail the possible sentence types in English: simple declarative sentences, such as "John ate the ice cream," passive sentences such as "The ice cream was eaten (by John)," and many others. Note that these sentences can be divided into constituent parts or "phrases" such as Subject and Predicate, and the Predicate into Verb and Object. We can bracket these pieces by the symbols [and]. Further, the Predicate itself might contain another Sentence, as in:

[Sentence [Subject John] [Predicate [Verb ate] [Object the ice cream-i [Sentence that [Subject Bill] [Predicate likes [Object x-i]]]]]]

which can be put into a more readable upside down tree structure that encodes the same information:

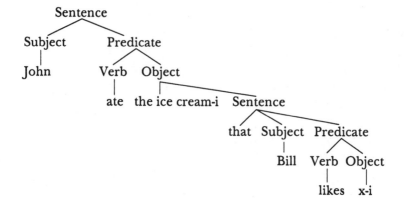

We have to unravel these pieces or "constituent structures" in order to understand what a sentence means—because the Subject ordinarily tells us who did the eating or liking, and the Object of the Predicate is the thing eaten or liked. This can be subtle. Here, for example, the Object of "like"—the thing liked—does not show up as pronounced where it usually does, right after "like" (as in "Bill likes John"). Instead the Object is really "the ice cream," and we have indicated this by placing an "x" with an index i right after "likes" and a corresponding index i after "ice cream." People must do something like this reflexively, without evident effort, for the most part. (If this does not seem overly complex, I will offer an even subtler example below that our computer system, like people, can and must handle.) To understand English, then, a computer must be at least as clever as this. Obviously, though, there are many more complex sentence types that make matters even harder: "It seems that Mary likes ice cream," "Who likes ice cream," and so forth— many tens of thousands for English, or so it seems.

Indeed, since sentences can go on and on and on, like so many in this essay, the first linguistic revolutionaries set out to capture at least part of our ability to create an "infinite variety of sentences from finite means." For a computer or a brain this is plainly a necessity, because no matter how cheap memory chips get, we cannot buy an infinite number of them, and no matter how big our heads get, even at MIT, they cannot get infinitely big. For at least 30 years, then, generative grammars have included a finite-rule component—a phrase structure grammar dictating that Sentences can contain other Sentences, ad infinitum, and so can recursively output all the possible Subject–Predicate constituent structures.

In addition, the first revolution also recognized that there was a systematic relationship between the constituent structures associated with sentences like "John ate the ice cream" and "The ice cream was eaten by John." They are related by a "transformational rule," in this case a Passive Rule. This Passive Rule takes as input a declarative sentence in the form Subject–Verb–Object and outputs a passive sentence by (1) interchanging the Subject and Object; (2) adding "be"; (3) changing the inflection of the verbs ("be" becomes "was"); and (4) adding a new structure, a phrase with "by" and the Subject in it. Note that using a rule already gives us more than we would get if we just listed the active and passive sentences separately, because it is predictive. Putting aside many complications, it says that wherever the active form is found, so will the passive, but not vice versa. Indeed this seems to be so, since even with "frozen" stock phrases like "kick the bucket," we have both the active "John kicked the bucket" and the passive "The bucket was kicked by John," but never just the passive form by itself. It is also clearly simpler and shorter, because we need not list out the passive forms anymore, but can generate them by rule.

Our story so far: The first linguistic revolution established a need for two sets of rules, one for basic phrases and the other for transformations, and together these aimed to cover all possible sentences. Each human language demanded different sets of rules—different possible basic constituent structures and transformational rules. Did this work for computers? Alas, not very well. For one thing, given a bare sentence like "The ice cream was eaten," the computer would have to make many guesses as to the possible basic or nonbasic constituent structures and the possible transformations that could have produced these forms as output—that is, it would have to "invert" the transformations. Since transformations could either add or delete words or phrases, many blind alleys would have to be explored. Here, for example, there is a "nothing"—silence—after "eaten" in the surface sentence (much like after "know" in "the girl that Bill knows"); because almost anything could replace a nothing, precious computer time is lost chasing down false leads.

Worse, extending the approach to specialized sublanguages revealed all sorts of exceptions that complicated the system's basic simplicity. Consider recipes. When Fanny Farmer says, "Place two cups of flour in a steel bowl. Add 1 cup sugar and butter. Whisk until blended," we note among other things that the last sentence does not mention the implied flour, sugar, and butter. Ordinarily, outside the recipe sublanguage, if we said just "John put," that would seem odd.

For this reason, among others, the first revolution in linguistic science did not find its way into commercially viable computer systems for using human languages. It fell into disfavor. Indeed, to this day, and even for the most advanced computer systems, it is currently fashionable, and even argued as sound engineering practice, to insist that there is no generative grammar at all in the sense described above, but rather a collection of many, many structural sentence patterns: passive sentences, active recipe sentences using "whisk," active recipe sentences using "fry," and so on and on, recursive only to handle the ever-present infinity we noted earlier in language. The theory's dirty laundry became laundry lists. Linguistic Picasso's optimistic rose period dissolved to a more morose blue.

Blue indeed, because the laundry-list approach comes close to saying there is no theory of human language at all in the conventional sense. It would be as if there were no atomic theory for chemistry but only lists of compounds. The disadvantages of such a situation should be clear: If true, we would have no short statement of the family resemblances between chemical compounds.

For language, we might dub this the "Tower of Babel Model" because it advocates a different "grammar" for each particular situation—different constituent structure patterns not only for English and Japanese, but also for recipes and medical texts, for lawyers, and, as in some Australian languages, for mothers-in-law. It would be as if we had two groups of people in separate isolation chambers, one preparing a rule list for English and the other for Japanese. Then there would be no reason at all to expect the English "I know what John bought" and the corresponding Japanese, "Watashi-wa Taro-ga nani-o katta ka shitte iru" (literally, "I John what bought Question-marker know past-tense") to be fundamentally alike. The two could be as wildly different as, well, Martian and Vulcan. Language acquisition would remain steeped in mystery. How could a child come equipped with all these different lists or even an ability to learn them all easily? Worse, from our computational perspective, we would have to reprogram our computer for each distinct language. What of our global language village now?

Not to worry. As we shall see, despite appearances, we can topple the Tower of Babel. The isolation chamber scenario plays to a false analogy, because people's language organs are not that divergent. While the "look and feel" of English and Japanese might seem quite different, remarkably the underlying language organ primitives are just four bits apart. Herein lies the promise for language and computation in the 21st century. It turns out that the first linguistic revolution three decades ago got the idea about sentence chemistry right; it just got the basic atoms wrong. It took a second, ongoing revolution at MIT, starting about 10 years ago, to get the primitives right—at least, on the right track. With the right atoms in hand, we can now build programs that handle English, Japanese, language acquisition, and even translation.

The New Language Chemistry

To see how the linguists discovered they were on the right track, we must don our chemistry hats again and recall the definition of an atomic element: something that, for the purposes of chemical combination, is not further divisible. Now let us turn to the the Passive Rule again. If this rule were a real atom, its components ought never to be seen on their own. What are these components? The Passive Rule includes movement of the object; inserting a "be" and changing the main verb's look so it ends in a form of "en"; and adding the "by" phrase. A decade or so ago, Noam Chomsky noted that all of these parts can occur on their own. The object can move to the Subject position without a verb, as in "the destruction of the city" becoming "the city's destruction" (notice that "the city's" is in the same position the Subject would be in if "destruction" were a Predicate); the "by" phrase, as in "the book understandable by everyone"; the Passive "en" verb form, as in "the melted ice cream." Further confirmation comes from examining languages that have different word orders than English but still exhibit passivelike sentences—languages such as Japanese, German, or Dutch, in which the verb comes at the end, revealing that even the Subject–Verb–Object Passive Rule triggering pattern was illusory. In Dutch we have "Kees zei dat Jan Marie kuste" ("Kees said that Jan Mary kissed") and the corresponding passive form "Kees zei dat Marie door Jan gekust werd" ("Kees said that Mary by Jan kissed was").[1]

Aha! But if the Passive Rule breaks down into further parts, it is not indivisible. Therefore, it is an epiphenomenal interaction of deeper principles, and these are the real atoms of human language. While we cannot retrace here the linguistic sleuthing needed to track down these fundamental principles, we can list some of them and note how simple they are. Figure 1, which shows the computer system that implements these principles, gives the entire lot of them along the right-hand side of its display. They comprise the full set of "rules" we use in our working system to analyze all languages, not just English, a remarkable fact considering the many thousands of rules the laundry-list approach adopted. Two crucial points follow. First, the joint interaction of just a handful principles replaces the effects of many thousands of rules. Second, some principles themselves are "parameterized" in that they can be switched in just two or three ways to yield the differences that we see from language to language. Together, then, this new approach to language, dubbed "principles and parameters theory," aims to solve the problems of language variation, acquisition, and processing through a radically different language chemistry. Here are some of these principles:

1. Languages allow just two basic tree shapes for their constituent phrases: (i) function–argument order, as in English where a verb begins a Predicate phrase, a preposition begins a Prepositional phrase, and so forth ("eat the ice cream," "with jimmies"), or (ii) the mirror image, argument–function form, as in Japanese or German.

2. Every verb must assign a thematic role to its "arguments" that says, roughly, who did what to whom, and every Subject and Object must receive exactly one such role. For instance, a main sentence with "eat" must mention the eater and implicitly the thing eaten, whereas "put" must mention the thing that is put somewhere and the location it is put ("John put the book" is odd, as is "John saw Mary Bill"). It should be clear that to know what "eat" means is to know at least this much.

3. Overt or pronounced Subjects or Objects must receive "Case," where by "Case" we mean an abstract version of what one would find in a Latin grammar. The Subject position receives, or is assigned, nominative case from the inflection of a verb; the Object of the verb receives accusative case; the object of a Preposition receives oblique case; and so on. (The ghost of this Latinate past haunts English but lightly, as in "she" vs. "her.") An adjective verb form such as "eaten" (compare "tired") does not assign Case. "Was" (like any so-called inflected verb) does assign Case. The parameterization remains minimal. In some languages, including English, Case assignment holds only under strict adjacency, which is why "John kissed quickly Mary" seems odd; in other languages, such as French, strict adjacency does not hold, which is why the equivalent French is fine.

4. A Subject or Object (in fact, any constituent) can move anywhere (subject to language-particular parameters). When it does, it leaves behind an invisible (unpronounced) element, a trace, linked to itself. (Note that this is much like the link we saw earlier between "ice cream" and the position after "likes" in "the ice cream that Bill likes.")

Can this really be all? Yes, indeed. Together these principles conspire to reproduce the effects not only of the basic Passive Rule but also of related forms like "the ice cream seems to have been eaten" that were not even covered by the original. The beauty of the new scheme is that now, at last, we can get computers to work with the system, without incurring the revenge of the Sagan numbers: "billions and billions" of rules.

But how? The picture is that possible sentences like "the ice cream was eaten" must "pass" the four required principles, now acting as constraints, where we understand the ice cream sentence to have been derived from the original form "was eaten the ice cream." It is as if the original form must "run a gauntlet" through a cascade of principles, emerging on the other side in its possible surface guise. In fact, we will see that this is roughly how our computer system operates, in reverse. In our example, since "eaten" acts like an adjective, it does not assign Case to the Object position immediately following the verb. Therefore, no pronounced Subject or Object had better follow—which is true. But "ice cream" must receive Case from somewhere and must also get a thematic role, namely the thing eaten. How? The only way is to move it to a position where it can get both. Principle 4 says that we can displace it, moving "ice cream" to Subject position and leaving an invisible trace after "eaten," where, as usual, the inflected verb "was" will assign nominative Case. In turn, the thematic role of the thing eaten, "ice cream," will be "transmitted" from its typical position after the verb in virtue of the trace left behind. (Note that if an ice-cream "eater" were originally present, we could move that to the object of a "by" phrase, where it too will get Case and a thematic role via the preposition "by," which assigns the thematic role of "agent" in conjunction with any verb in such a context. The by-phrase and the eater need not appear, however.)

So all is well. The sentence jumps all the hurdles, and we never had to mention a Passive Rule. The "passive" sentence is thrust upon us, an inevitable consequence of the atomic structure of language and a few particulars of English. (Observant readers may ask why we just do not place "ice cream" as the object of the by-phrase; then it would get Case and a thematic role. That is right, but two problems bar this way out: First, "ice cream" would not get the right thematic role, since it will become the eater, not the thing eaten. Second, this would violate a constraint not mentioned earlier that applies to moved constituents: At the very least, the moved phrase cannot wind up lower down in the tree structure than the trace left behind.)

This may seem like a lot of pain for little gain. It is, for a single example. The big gain comes in the fact that, once mastered, these principles can be recombined in different ways to cover a myriad of other sentence types. For instance, to handle a sentence like "the ice cream seems to have been eaten," all we have to note is that the Subject position of "seems" is actually vacant and ready to receive a displaced ice cream—as we can tell from the dummy "it" in examples like "It seems that John ate the ice cream." There is no thematic role assigned to this position by "seems"—which is suspiciously like the "was eaten" sentences. So we should expect a sentence type in which "ice cream" can move to this position; and what we expect, we get. Similarly for the Dutch examples cited earlier: None of the forcing principles even mention verb position, so we expect, and find, the corresponding Dutch passive-type sentences without having to say anything new at all. That is the great economy of the new atomic theory. In addition, we now have a budding account of language acquisition: If all that differs from English to German to Japanese are the settings of a handful of parameters, then there really is very little for the child to learn. Hearing English, the child sets the "tree branching parameter" in one direction; hearing Japanese, she sets it in the other direction.

Babel Undone: The New Marriage of Computation and Linguistics

We can now assemble a computer system that incorporates these principles directly. In effect, just as the new atomic theory says that there is only a single human language, so we need only a single set of principles, as shown in in the right-hand column of the computer screen snapshot in figure 1.[2] Here are the full 20-odd principles that we need to cover not only English (as shown here) but also Japanese (as shown in figure 3). What we called principle 1, tree branching left or right, is the first box labeled "Parse S-Structure" under the "Generators" column about two-thirds of the way down the right-hand side; principles 2 and 3 are labeled as such; principle 4, movement, is called "Move alpha." Take note: The principles are the same in both languages—besides the different words, the only clues that the machine is processing English in figure 1 and Japanese in figure 2 are the deliberately placed "E" and "J" in the upper right-hand corners of the screens.

Even so, with just a handful of principles, quite subtle sentences can be tackled. Figure 1 shows our implemented system in action

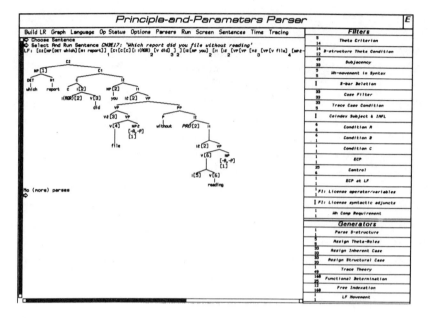

Figure 1

analyzing "Which report did you file without reading?" People tacitly, but easily, overcome a number of complexities to understand what such a sentence means. It is a question about what was filed (the Object of "file"); importantly, whatever report was filed is the same report that was read or not. Further, whoever "you" is, that same "you" is the Subject of "reading." In short, the sentence can be construed something like, "for which report X did you file the report X without (you) reading the report X." So there are three invisible, "unpronounced" items—the Object of "file," the Subject of "reading," and the Object of "reading"—and "which report" is linked to two of them. Now, such a sentence would give even the laundry-list approach fits, for it is ordinarily set up to link a single question phrase like "which report" to a single invisible Object, but not to two. A special-purpose rule would have to be written to deal with this case.

No such special-purpose rules are needed in figure 1. We can see the resulting bracket and tree structure outputs (preceded by the label "LF:") and the thematic roles assigned in the middle of the screen. The very same principles that yield active and passive sentences, simple questions, and so forth also admit the sentence shown. Astonishingly, nothing new need be said at all. This complex sentence is just another possible "chemical compound" derivable from the basic linguistic atoms and their combinatory constraints.

Figure 1 also suggests how our computer processor works. As hinted earlier, the computer essentially runs the principle gauntlet in reverse. We start with the bare sentence. Using its knowledge about the basic tree branching structure for the language, the system guesses an initial constituent phrase scaffolding on which to hang the remaining principles. Each principle in turn tries its hand at doing some chemistry, either attempting to add some structure or winnowing out sentence "compounds" that other principles may have proposed. Finally, one or more structures emerge from the gauntlet, as shown in the middle of the screen.

Lacking a videotape, we can only imagine the computer at work, because the "compounds" are actually swung back and forth between the principles; the system does not generate all possible compounds before proceeding to the next principle in line. Still, we can get some feel for the action by observing in the figure how each principle comes with a paired set of numbers, top and bottom. These indicate the possible "chemical compounds"—the structures—that were passed to each principle, with the top number being the possibilities going into the principle and the bottom number being the possibilities that emerge unscathed. For instance, in our example we can see our brave sentence diving into the initial tree-structure scaffolding via the "Parse S-structure" box underneath the larger label "Generators." Observe that one sentence goes in—the top number—and one possibility, one structure, comes out. This structure is then passed to the movement principle box, Move alpha, and 49 new "compounds" come out (since we can move anything anywhere, this should be no surprise). At last the shuffling back and forth dies down, and one winner emerges from the pack: the analysis that people get for "What report did you file without reading," as shown with brackets and trees. All this takes but a second or two, for many dozens of sentence types, as figure 2 shows. In the best MIT tradition, then, we have married the finest scientific theory of language we know to good engineering, producing a superior computer system for processing English. No rules need apply.

The best is yet to come. Since the linguistic theory says from the start that there is, in effect, but one human language, we do not have too much to change at all to build a system for analyzing Japanese. Japanese is "almost" like English—all we have to do is

1. Someone likes everyone
2. Who that John knows does he like
3. He likes everyone that John knows
4. It is likely that John is here
5. I am eager for John to be here
6. I believe John to be here
7. I believe John is here
8. I want to be clever
9. John was persuaded to leave
10. John was arrested by the police
11. I believe John to be intelligent
12. John was believed to be intelligent
13. I am proud of John
14. I wonder who you will see
15. Their pictures of each other are nice
16. John likes Mary's pictures of him
17. Who does he think Mary likes
18. The men think that pictures of each other will be on sale
19. The men think that Mary's pictures of each other will be on sale
20. Which report did you file without reading
21. The report was filed without reading
22. John is certain to see this
23. Who do you think that John saw
24. Who do you think John saw
25. Who do you think saw Bill
26. Who will read what
27. Why did you read what
28. Who believes the claim that Mary read what

Figure 2. Some sentence types handled by the system.

throw four switches one way rather than another. To be sure, we must also supply a new dictionary, but that is understandable since there is no way our system, or most people, could know that "kaimashita" means "buy" in Japanese, or what roles it assigns. Beyond this, though, it cannot be emphasized strongly enough that nothing else need be said. The same algorithms and the same principles can be used, as we can see in figure 3, where the computer analyzes our earlier example, "Watashi-wa Taro-ga nani-o katta ka shitte iru" ("I know what John bought"). We did not need thousands of new rules for sentences like these, or for even more

Electronic Organs of the Mind 115

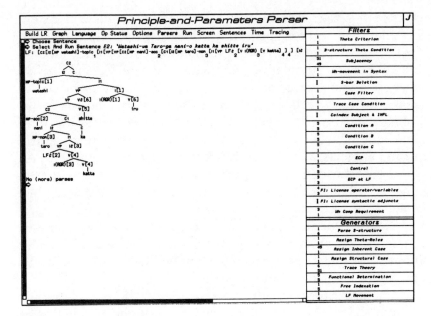

Figure 3

complex examples. In fact, we did not need to add any new rules at all. No handcrafted reprogramming was required, other than these few statements:

1. Japanese sentence structures branch to the left (see figure 3). English sentence structures branch to the right (see figure 1). (Japanese Verbs are at the end of Predicates; Prepositions are "Postpositions," at the end of their phrases.)

2. Words like "why" ("naze") or "what" ("katta") cannot move overtly, and so "I know what John bought" winds up spoken as roughly "I know John bought something, " even though these words do move in Japanese speakers' mental computations to determine meaning, since the Japanese question means the same thing as the English. In English, these words or phrases can move overtly.

3. Japanese can drop Subjects freely. English cannot. (Compare the impossible English "left, " for "I left," with Spanish or Italian, which can also drop Subjects, as in "parlo italiano," meaning "I speak Italian.")

4. Case adjacency need not hold for Case assignment in Japanese. English requires Case adjacency.

Taken together, parametric differences 1–4 plus a different dictionary account for apparently vast surface variations between English and Japanese: first, the verb at the end of Japanese sentences; second, the relatively free word order or "scrambling" in Japanese—sentences often begin with a topic, but often objects can appear in front of the subject, as in "John Mary a book buy" or "John-ga Mary-ni hon-o ageta," which could appear as "a book John Mary bought" or "hon-o John-ga Mary-ni ageta," and so forth; third, missing Subjects, as in "Pen-o kaimashita" (He bought a pen). All of this goes through the computer system as before, as can be seen in figure 3: The lack of Case adjacency allows phrases to scramble; "what"-type words stay fixed but can move in one's mental representation in order to fix meaning; the trees branch the other way. The language changes; the principles stay the same. The new linguistic atomic theory works, and we at MIT have built the world's first computer system that can use it.

As Chomsky observes, we may "conclude that there is really only one human language, with some marginal differences" and that "it may be that computational systems are virtually identical for all languages."[3] What the engineers and the linguistic scientists have done, joining hands across Vassar Street, is to build an electronic version of that computational system, an electronic organ of the mind. The abstract "maybe" can be turned into a much stronger, concrete reality—a reality that at least has the potential to bind the world closer together by recognizing both the common, special ability that makes us all human and the computational engines that wire us together. Is this second revolution worth a new plaque in the Kendall Station subway time line? Perhaps not only in Kendall Square, but also in the subways of Tokyo, Berlin, Moscow, and Paris—wherever the marvel of human language can be found.

Acknowledgments

The research reported on in this essay was supported in part by a John Simon Guggenheim Fellowship, a Presidential Young Investigator Award, and a grant from the Kapor Family Foundation to Robert C. Berwick. The design and implementation of the computer system described here was carried out as Sandiway Fong's Ph.D. dissertation under the direction of Professor Berwick, and is presented in full detail in their forthcoming book from MIT Press. This research could not have been carried out without the strong interdisciplinary support that makes MIT thrive, in particular, the intellectual stimulation of the MIT Department of Linguistics and Philosophy and Professors Noam Chomsky, Morris Halle, Samuel Jay Keyser, Kenneth Hale, David Pesetsky, and Alec Marantz.

Notes

1. All the sentence examples in this paragraph are from H. Kolb and C. Thiersch, "Levels and empty categories in a principles and parameters approach to parsing," in H. Haider and K. Netter, eds., *Representation and Derivation in the Theory of Grammar* (Dordrecht: Kluwer), 252–253.

2. The computer implementation and screen snapshots are from Sandiway Fong, "Computational properties of principle-based grammatical theories," Ph.D. dissertation, Department of Electrical Engineering and Computer Science, MIT.

3. Noam Chomsky, "Language from an internalist perspective." unpublished manuscript, MIT Department of Linguistics and Philosophy, 1991.

Further Reading

R. Berwick, S. Abney, and C. Tenny, *Principle-Based Parsing* (Dordrecht: Kluwer, 1991).

R. Berwick and S. Fong, "Principle-based parsing: Natural language parsing for the 1990s," in P. Winston and S. Shellard, eds., *Artificial Intelligence: Expanding Frontiers* (Cambridge: MIT Press, 1990).

R. Berwick and S. Fong, *Computation with Principle-Based Linguistic Theories* (Cambridge: MIT Press, 1992).

System Dynamics: Adding Structure and Relevance to Precollege Education
Jay W. Forrester

Secondary education in the United States is under increasing attack for not preparing students to cope with modern life. Failures appear in the form of corporate executives who misjudge the complexities of growth and competition, government leaders who are at a loss to understand economic and political change, and a public that supports inappropriate responses to immigration pressures, changing international conditions, rising unemployment, the drug culture, governmental reform, and inadequacies in education.

Growing criticism of U.S. education may focus attention on incorrect diagnoses and ineffective treatments. Weakness in education arises not so much from poor teachers as from inappropriateness of material that is being taught. Students are stuffed with facts without having a frame of reference for making those facts relevant to the complexities of life. Responses to educational deficiencies are apt to result in public demands for still more of what is not working. Pressures will increase for additional science, humanities, and social studies in an already overcrowded curriculum. Instead, an opportunity exists for moving toward a common foundation that pulls all fields of study into a more understandable unity.

Sources of Educational Ineffectiveness

Much of the current dissatisfaction with precollege education arises from past inability to show things whole, to convey how people and nature interact, and to reveal causes for what students see happening. Because of its fragmentary nature, traditional education becomes less relevant as society becomes more complex, crowded, and tightly interconnected.

Education is compartmentalized into separate subjects that, in the real world, interact with one another. Social studies, physical science, biology, and other subjects are taught as if they were inherently different from one another, even though behavior in each rests on the same underlying concepts. For example, the dynamic structure that causes a pendulum to swing is the same as the core structure that causes employment and inventories to fluctuate in a product-distribution system and in economic business cycles. Humanities are taught without relating the dynamic sweep of history to similar behaviors on a shorter time scale that a student can experience in a week or a year.

High schools teach a curriculum from which students are expected to synthesize a perspective and framework for understanding their social and physical environments. But that framework is never explicitly taught. Students are expected to create a unity from the fragments of educational experiences, even though their teachers have seldom achieved that unity.

Missing from most education is a direct treatment of the time dimension. What causes change from the past to the present and the present into the future? How do present decision-making policies determine the future toward which we are moving? How are lessons of history to be interpreted to the present? Why are so many corporate, national, and personal decisions ineffective in achieving intended objectives? Conventional educational programs seldom reveal the answers. Answers to such questions about how things change through time lie in the dynamic behavior of social, personal, and physical systems. Dynamic behavior, common to all systems, can be taught as such. It can be understood.

Education has taught static snapshots of the real world. But the world's problems are dynamic. The human mind grasps pictures, maps, and static relationships in a wonderfully effective way. But in systems of interacting components that change through time, the human mind is a poor simulator of behavior. Mathematically speaking, even a simple social system can represent a tenth-order, highly nonlinear, differential equation. Mathematicians cannot solve the general case for such an equation. No scientist, citizen, manager, or politician can reliably judge such complexity by intuition. Yet, even a junior high school student with a personal computer and coaching in computer simulation can advance remarkably far in understanding such systems.

Education faces the challenge of undoing and reversing much that people learn by observing simple dynamic situations. Experiences in everyday life deeply ingrain lessons that are deceptively misleading when one encounters more complex social systems.[1] For example, from burning one's fingers on a hot stove one learns that cause and effect are closely related in both time and space. Fingers are burned here and now when too close to the stove. Almost all understandable experiences reinforce the belief that causes are closely and obviously related to consequences. But in more complex systems, the cause of a difficulty is usually far distant in both time and space. The cause originated much earlier and arose from a different part of the system from where the symptoms appear.

To make matters even more misleading, a complex feedback system usually presents what we have come to expect, an apparent cause that lies close in time and space to the symptom. However, that apparent cause is usually a coincident symptom through which little leverage exists for producing improvement. Education does little to prepare students for succeeding when simple, understandable lessons so often point in exactly the wrong direction in the complex real world.

Cornerstones for a More Effective Education

Two mutually reinforcing developments now promise a learning process that can enhance breadth, depth, and insight in education. These two are system dynamics and learner-directed learning.

Precursors of System Dynamics

System dynamics evolved from prior work in feedback-control systems. The history of engineering servomechanisms reaches back several hundred years.[2] In the 1920s and 1930s, understanding the dynamics of control systems accelerated. New theory evolved during development of electronic feedback amplifiers at the Bell Telephone Laboratories and during work at MIT on feedback controls for analog computers and military equipment.

The present thread leading to system dynamics started when I was introduced to feedback systems by Gordon S. Brown in the MIT Servomechanisms Laboratory in the early 1940s. After 1950, people became more aware that feedback control applies not only to engineering systems but also to all processes of change—biological, natural, environmental, and social.

System Dynamics in Precollege Education

During the last 30 years, those in the profession of system dynamics have been building a more effective basis than previously existed for understanding change and complexity. The field rests on three foundations:

1. Growing knowledge of how feedback loops—containing information flows, decision making, and action—control change in all systems. Feedback processes determine stability, goal seeking, stagnation, decline, and growth. Feedback systems surround us in everything we do. A feedback process exists when action affects the condition of a system and that changed condition affects future action. Human interactions, home life, politics, management processes, environmental changes, and biological activity all operate on the basis of feedback loops that connect action to result to future action.

2. Digital computers, now primarily personal computers, to simulate the behavior of systems that are too complex to attack with conventional mathematics, verbal descriptions, or graphical methods. High-school students, using today's computers, can deal with concepts and dynamic behavior that only a few years ago were restricted to work in advanced research laboratories. Excellent user-friendly software is now available.[3]

3. Realization that most of the world's knowledge about dynamic structures resides in people's heads. The social sciences have relied too much on measured data. As a consequence, academic studies have failed to make adequate use of the database on which the world runs—the information gained from living experience, apprenticeship, and participation. Junior-high and high-school students already have a vast amount of operating information about individuals, families, communities, and schools from which they can learn about social, business, economic, and environmental behavior.

The system dynamics approach has been successfully applied to behavior in corporations, internal medicine, fisheries, psychiatry, energy supply and pricing, economic behavior, urban growth and decay, environmental stresses, population growth and aging, training of managers, and education of primary- and secondary-school students.

Nancy Roberts first demonstrated system dynamics as an organizing framework at the fifth and sixth grade levels.[4] Her work showed the advantage of reversing the traditional educational sequence that normally progresses through five steps:

1. learning facts,
2. comprehending meaning,
3. applying facts to generalizations,
4. analyzing to break material into constituent parts,
5. synthesizing to assemble parts into a whole.[5]

Most students never reach that fifth step of synthesis. But, synthesis—putting it all together—can be placed near the beginning of the educational sequence. By the time students reach junior high school they already possess a wealth of facts about family, interpersonal relationships, community, and school. They are ready for a framework into which the facts can be fitted. Unless that framework exists, teaching still more facts loses significance.

In his penetrating discussion of the learning process, Bruner states, "the most basic thing that can be said about human memory... is that unless detail is placed into a structured pattern, it is rapidly forgotten."[6] For most purposes, such a structure is inadequate if it is only a static framework. The structure should show the dynamic significance of the detail—how the details are connected, how they influence one another, and how past behavior and future outcomes arise from decision-making policies and their interconnections.

System dynamics can provide that framework to give meaning to detailed facts. Such a dynamic framework provides a common foundation beneath mathematics, physical science, social studies, biology, history, and even literature.

Despite the potential power of system dynamics, it could well be ineffective if introduced alone into a traditional educational setting in which students passively receive lectures. System dynamics can not be acquired as a spectator sport any more than one can become a good basketball player by merely watching games. Active participation instills the dynamic paradigm. Hands-on involvement is essential to internalizing the ideas and establishing them in one's own mental models. But traditional classrooms lack the intense involvement so essential for deep learning.

Learner-Directed Learning

Those who have experienced the excitement and intensity of a research laboratory know the involvement accompanying new discoveries. Why should not students in their formative years experience similar exhilaration from exploring new challenges? That sense of challenge exists when a classroom operates in a "learner-directed-learning" mode.

Listening to lectures presents students with a deadening, nonparticipating, undemocratic, authoritarian process. It has the disadvantages we normally associate with authoritarian governments. The recipients of such lectures naturally resist authority, they sabotage the process, and their rebellion defeats the best intentions of educators.

Learner-directed learning, is a term I first encountered from Mrs. Kenneth Hayden of Ideals Associated.[7] It substantially alters the role of a teacher. A teacher is no longer a dispenser of knowledge addressed to students as passive receptors. Instead, where small teams of students work together to help one another, a "teacher" becomes a colleague and participating learner. Teachers act as guides and resource persons, not as authoritarian figures dictating each step of the educational process. The relationship is more like being a thesis adviser than a lecturer.

Perhaps the best way to glimpse the combination of system dynamics and learner-directed learning is to hear from a few of those having first-hand experience. After the Servomechanisms Laboratory, Gordon Brown became head of the MIT Electrical Engineering Department and then Dean of Engineering before retiring in 1973. In the late 1980s, he completed the circle he had originally launched by picking up system dynamics and introducing it into the Orange Grove Junior High School in Tucson, Arizona. He describes his role as the "citizen champion" engaged in drawing all participants in the school system together in their search for a new kind of education:

The use of computers in the classroom (not in a computer lab) has, for us in Tucson, resulted in a very unique learning environment... (students) learn what they need to know as the teacher guides them in conducting a simulation in class. They work in groups, two or three to a computer—certainly not one per computer—and thereby help one another. Dr. Barry Richmond says that this situation, in effect, multiplies the number of teachers by the number of students. Before doing a simulation the students spend several class periods gathering information about the topic; they take notes during lectures, learn about a library and read references, and, working as a group, plan the simulation. By working this way Draper's students do not merely try to remember the material for a test but actually have to use it in a project simulating real life situations. This has led us to identify a new teaching paradigm which we define as SYSTEM THINKING with LEARNER DIRECTED LEARNING.[8]

Gordon Brown introduced system dynamics to Frank Draper, eighth-grade biology teacher, by loaning him a computer and software for a weekend. Draper returned with the comment, "This is what I have always been looking for, I just did not know what it might be." At first, Draper expected to use system dynamics and computer simulation in one or two classes during a term. Then he found they were becoming a part of every class. With so much time devoted to new material, he feared he would not have time to cover all the required biology. But, two-thirds of the way through the term, Draper found he had completed all the usual biology content. The more rapid pace had resulted from the way biology had become more integrated and from the greater student involvement resulting from the systems viewpoint. Also, much credit goes to the "learner-directed learning" organization of student cooperative study teams within the classroom. To quote Draper, "There is a free lunch." He writes of his classroom experience:

Since October 1988 our classrooms have undergone an amazing transformation. Not only are we covering more material than just the required curriculum, but we are covering it faster (we will be through with the year's curriculum this week and will have to add more material to our curriculum for the remaining 5 weeks) and the students are learning more useful material than ever before. 'Facts' are now anchored to meaning through the dynamic relationships they have with each other. In our classroom students shift from being passive receptacles to being active learners. They are not taught about science per se, but learn how to acquire and use knowledge (scientific and otherwise). Our jobs have shifted from dispensers of information to producers of environments that allow students to learn as much as possible.

We now see students come early to class (even early to school), stay after the bell rings, work through lunch and work at home voluntarily (with no assignment given). When we work on a systems project—even when the students are working on the book research leading up to system work—there are essentially no motivation/discipline problems in our classrooms.[9]

A dynamic framework can even organize the study of literature as described by Pamela Hopkins. She teaches high-school juniors in a slower-track group where few of the students had shown even a slight interest in anything like a play by Shakespeare:

[When we used] a STELLA model which analyzed the motivation of Shakespeare's Hamlet to avenge the death of his father in *Hamlet*... The students were engrossed throughout the process... The amazing thing was that the discussion was completely student dominated. For the first time in the semester, I was not the focal point of the class. I did not have to filter the information from one student back to the rest of the class. They were talking directly to each other about the plot events and about the human responses being stimulated. They talked to each other about how they would have reacted and how the normal person would react. They discussed how previous events and specific personality characteristics would affect the response to each piece of news, and they strove for precision in the values they assigned for the power of each event. My function became that of listening to their viewpoints and entering their decisions into the computer. It was wonderful! It was as though the use of precise numbers to talk about psychological motives and human responses had given them power, had given them a system to communicate with. It had given them something they could handle, something that turned thin air into solid ground. They were directed and in control of learning, instead of my having to force them to keep their attention on the task.[10]

The students in Hopkins's class were from an underprivileged section of the city and most had been labeled as poor students. Several months after the experience related above, I received a letter from Louise Hayden, director of Ideals Associated:

Pam and I are so pleased and surprised at the ongoing involvement and depth of interest the high school students in her workshop of last June are showing. They are meeting with her weekly after school, eager to learn more about system dynamics and to use their advances to help a younger student learn. They are arousing considerable teacher interest as they try to use causal loops in all their classrooms. Information is flowing upward—and from students who varied in achievement from high to very low.

We attribute the enthusiasm and commitment to their sense of the potential of system thinking, and to the feelings of self-worth from being regarded as educational consultants. It is their first experience in learner-directed learning. This may well be the first time they have considered themselves a responsible part of the social system.[11]

Many people assume that only the "best" students can adapt to the style of education here suggested. But who are the best students? Results so far indicate no correlation between students who do well in this program and how they had been previously labeled as good or poor students. Some of the so-called poor students find traditional education lacks relevance. They are not challenged. In a different setting they come into their own and become leaders. Some of the students previously identified as best are strong on repeating facts in quizzes but lack an ability to synthesize and to see the meaning of their facts. Past academic record seems not to predict how students respond to this new program.

The Present Status

System dynamics is developing rapidly, but does not yet have widespread public visibility. The international System Dynamics Society was formed in 1985. Membership has grown to some 300. Annual international meetings have been held for 15 years in locations as widely spread as Norway, Colorado, Spain, China, California, Germany, and Thailand. System dynamics books and papers are regularly translated into many languages including Russian, Japanese, and Chinese.

After 30 years of development, several dozen books present the theory, concepts, and applications of system dynamics. Some have exerted surprising public impact.[12] The *Limits to Growth* book, showing interplay among population, industrialization, hunger, and pollution, has been translated into some 30 languages and has sold over 3 million copies.[13] Such widespread readership of books based on computer modeling testifies to a public longing to understand how present actions lead into the future.

Early leaders in system dynamics were educated at MIT, but competence is now appearing in many places. Talent exists on which to build a new kind of education, even though system dynamics is so broadly applicable throughout physical, social, biological, and political systems that the present small number of experts are thinly dispersed over a wide spectrum of activities.

Initial work with precollege schools is under way in several places. Through the efforts of Barry Richmond and others, system dynamics is now becoming established in some 30 junior and senior high schools. Several hundred schools have started preliminary exploratory activity. Progress is only now reaching a point where self-sustaining momentum exists in the absence of strong inputs from a person broadly knowledgeable about system dynamics.

Part of the educational emphasis will focus on "generic structures." A rather small number of relatively simple structures appear repeatedly in different businesses, professions, and real-life settings. Students can transfer insights from one setting to another. For example, one of Draper's eighth-grade students grew bacteria in a culture dish, then looked at the same pattern of environmentally limited growth through computer simulation. From the computer, the student looked up and observed, "This is the world population problem, isn't it?" Such transfer of insights from one setting to another will help to break down barriers between disciplines. It means that learning in one field becomes applicable to other fields.

There is now promise of reversing the trend of the last century toward ever greater fragmentation in education. There is real hope of moving back toward the "Renaissance man" idea of a common teachable core of broadly applicable concepts. We can now visualize an integrated, systemic, educational process that is more efficient, more appropriate to a world of increasing complexity, and more supportive of unity in life.

Several high schools, curriculum-development projects, and colleges are using a system dynamics core to build study units in mathematics, science, social studies, and history. These programs have not yet reached the point of becoming fully integrated educational structures.

The most advanced United States experiment in bringing system dynamics and learner-directed learning together into a more powerful educational environment appears to be in the Catalina Foothills School District of Tucson, Arizona. In that community the necessary building blocks for successful educational innovation have come together. Progress in that school system rests on:

1. fundamental new concepts of education,

2. a receptive community,

3. talented teachers who are willing to try unfamiliar ideas and who are at ease in the nonauthoritarian environment of learner-directed learning,

4. an understanding and encouraging school administration,

5. a supportive school board, and

6. a "citizen champion" who, without a personal vested interest in the outcome—except for a desire to facilitate improvement in education—has helped by inspiring teachers, finding funding, arranging for computers, and, above all, facilitating convergence of political processes in the community.

The Catalina Foothills district does not have its own high school. Students go into the Greater Tucson system. After seeing the impact on several hundred students of the new educational philosophy embedded in the Orange Grove junior high school, parents became reluctant to have children revert to a traditional high school. The District in 1990 voted a $30 million bond issue to create a high school in the educational pattern that had been pioneered in the junior high school.

Other countries (Norway, Germany, Japan, and China) are moving ahead rapidly in using system dynamics as a foundation for designing a powerful educational system below the college level. The Scandinavian countries are working together to build precollege education on a system dynamics foundation. I have received a German book detailing the experimental use of system dynamics and the STELLA software for teaching high-school physics. Even though this new basis for education was first developed in the United States, other countries in Europe and the Orient are moving at least as quickly to apply the ideas.

The Future

Over the next several decades, an improved kind of education can evolve. The growing severity of corporate, economic, social, political, and international difficulties demonstrates need for better understanding. Aggressive action can lead sooner to a society with keener insights into reasons for current shortcomings.

The basis now exists for a far more effective educational program. But a vast amount of work remains to build on the present foundation. Adequate educational materials are yet to be developed. One book exists aimed especially at high schools.[14] Although not written specifically for precollege use, other introductory system dynamics books are available.[15] Nevertheless, the published material does not yet adequately convey the background, simulation models, related teacher-support materials, and guidance on teaching methods. Much such material already exists in places ranging from files at MIT, to work of teachers who are pioneering in systems thinking and learner-directed learning. But most existing material is not now accessible.

Those who are developing system dynamics in secondary schools have of necessity been creating their own educational materials. No network has existed before 1991 for interchanging information among innovators in precollege education. But that missing link is now being remedied by a new office, the Creative Learning Exchange, established by John R. Bemis, to receive, print, and distribute system dynamics educational materials. That office will maintain communications between schools, encourage training seminars for teachers, advise teachers in preparing new materials for wider dissemination, and assist in maintaining the integrity and practicality of the system dynamics content of emerging curricula.

John Bemis also supports an active group of MIT students in the Undergraduate Research Opportunities Program (UROP), who are developing educational materials for use in high schools. The UROP students are working with teachers in the Cambridge Rindge and Latin High School to test materials and acquire experience in the real world of teachers and classrooms.

Sufficient success has been achieved to justify continued aggressive work to create a far more effective kind of precollege education. Private individuals are moving ahead, rather than waiting for public political organizations to innovate. For example, James L. and Faith Waters have made perceptive and timely funding available to encourage system dynamics in precollege education. Private support can operate with a freedom and a dedication to purpose that is seldom possible with the bureaucratic processes of government.

The next step is to establish a limited network of schools (starting with a few and expanding as progress justifies) that are experimenting with how best to introduce system dynamics and learner-directed learning into classrooms. Preliminary results are so intriguing that schools without the necessary ingredients for success may begin, then fail, and discredit the philosophy being presented here.

The next goal is to reach a point where about a dozen schools have been unambiguously successful and have achieved self-sustaining momentum. Thus far, the few dozen schools that are making good progress are relying on outside guidance to assist when barriers are encountered. Some are beginning to emerge from such dependence on external assistance, but there are not yet sufficient examples of ongoing, independent successes to overshadow failures that are almost certain to occur.

Solid foundations must be established. Curriculum materials are still insufficient and inadequate. The politics and processes of moving from a traditional school to a radically different style of education must be better understood. No one yet knows what percentage of present teachers can make the transition from traditional teacher-dominated classrooms to the freewheeling, research atmosphere of a learner-directed classroom. To some teachers, the transition is threatening. Little is known about how to evaluate students coming out of this different kind of education. Standardized evaluation may not be desirable or possible in a program that emphasizes individual development and diversity.

Creating a new kind of education will take substantial time. Planning and funding should provide for long-run continuity based on step-by step progress. Funding will be needed for publishing, retraining teachers, and launching demonstration schools.

Plans must provide a time horizon long enough for carrying through a radically improved process of education. In addition to a core of experts in system dynamics, there must be cohesive groups that understand the interrelated aspects of successful educational innovation. Such groups must combine experienced teachers, who understand the problems and opportunities in class rooms, with those who can translate ideas into effective teaching materials. "Citizen champions" can serve an important role in drawing together teachers, school administrators, school boards, parents, concerned public, and state and national school officials. Such influential groups are beginning to coalesce around the combined concepts of system dynamics and learner-directed learning.

Notes and References

1. Jay W. Forrester, "Counterintuitive Behavior of Social Systems," *Technology Review*, 73:3 (1971), 53-68.

2. George P. Richardson, *Feedback Thought in Social Science and Systems Theory* (Philadelphia: University of Pennsylvania Press, 1991).

3. High Performance Systems, *STELLA II Users Guide* (Hanover, NH: High Performance Systems, 1990); Alexander L. Pugh III, *Professional DYNAMO Plus Reference Manual* (Cambridge: Pugh-Roberts Associates, 1986). For most work at the precollege level, STELLA™ on Macintosh computers is easiest to use. It includes an excellent manual with learning exercises and an introduction to the philosophy of system dynamics. For more advanced professional use, DYNAMO™ is available for IBM and compatible computers. Several other software packages exist for system dynamics modeling, some with special attention to use in secondary schools.

4. Nancy Roberts, "A Dynamic Feedback Approach to Elementary Social Studies: A Prototype Gaming Unit," Ph.D. thesis, Boston University, 1975.

5. Nancy Roberts, "Teaching Dynamic Feedback Systems Thinking: an Elementary View," *Management Science*, 24:8 (1978), 836-843.

6. Jerome S. Bruner, *The Process of Education* (New York: Vintage Books, 1963), 24.

7. Ideals Associated in Tucson is a small foundation that for two decades has fostered an approach to learning that enlists students in an active participation that contributes to the momentum of the educational process.

8. Gordon S. Brown, "The Genesis of the System Thinking Program at the Orange Grove Middle School, Tucson, Arizona," personal report, March 1, 1990.

9. Frank Draper, personal communication, May 2, 1989.

10. Pamela Lee Hopkins, "Classroom Implementation of STELLA to Illustrate *Hamlet*," description of computer model and classroom experience, Desert View High School, Tucson, AZ, 1990.

11. Louise Hayden, personal communication, 1990.

12. Jay W. Forrester, *Urban Dynamics* (Cambridge: Productivity Press, 1969), and *World Dynamics*, 2d ed. (Cambridge: Productivity Press, 1973).

13. Donella H. Meadows et al., *The Limits to Growth* (New York: Universe Books, 1972).

14. Nancy Roberts et al., *Introduction to Computer Simulation: A System Dynamics Modeling Approach* (Reading: Addison-Wesley, 1983).

15. Jay W. Forrester, *Industrial Dynamics* (Cambridge: Productivity Press, 1961), *Principles of Systems*, 2d ed. (Cambridge: Productivity Press, 1968), *Urban Dynamics*; and *Collected Papers of Jay W. Forrester* (Cambridge: Productivity Press, 1975); Michael R. Goodman, *Study Notes in System Dynamics* (Cambridge: Productivity Press, 1974); George P. Richardson and Alexander L. Pugh III, *Introduction to System Dynamics Modeling with DYNAMO* (Cambridge: Productivity Press, 1981).

Clean Utilization of Fossil Fuels
János M. Beér

There is little doubt that, for the forseeable future, fossil fuels will remain our major energy resource and that their combustion will remain the most prevalent mode of energy utilization. Combustion is, however, a polluting process, and clean combustion technologies are needed for compliance with continually tightening emissions regulations. MIT is very much involved in creating these new technologies, and a review of current work in this area seems an appropriate offering for this volume.

The options for improving combustion technology fall into three main categories: (1) cleaning the fuel (e.g., coal beneficiation), (2) modifying the combustion process and creating new combustion processes, and (3) cleaning postcombustion flue gas (sulfur scrubbers, catalytic denitrification, etc.). Combustion process modification is probably the most technically challenging and yet the least expensive of these alternatives. The most cost-effective approach to emissions reductions is therefore to start by improving the combustion process and then to use flue gas treatment to further reduce pollutant emissions.

Combustion research has a long tradition at MIT. Hoyt C. Hottel founded the Fuels Research Laboratory in 1928 and served as its director until his retirement in 1968. In these four decades, the laboratory made important contributions to the maturation of industrial furnace design from an art to a science. The research level has continued to expand and presently includes faculty members Glenn C. Williams, Adel F. Sarofim, Jack B. Howard, John P. Longwell, and János M. Beér.

In 1976, MIT provided funding and space for two major research facilities. A 759 kilowatt fluidized combustor has been used to study fundamental aspects of this relatively young combustion technology. The major unit—a 3 megawatt (thermal) combustion tunnel known as the Combustion Research Facility (CRF)—was designed to permit detailed experimental investigations of turbulent flames in thermal environments that simulate conditions in large-scale furnaces.

The main objectives of current research are to elucidate the processes of formation and destruction of pollutants in flames and the mechanisms of transformation and deposition of mineral matter in boiler plants. These objectives are being pursued through a combination of bench-scale studies, computer modeling, and detailed experiments on a pilot scale. The chemical-physical rate data that are put into the models are obtained from the bench-scale studies, while the data on mixing in turbulent flames and model validation come from the combustion tunnel experiments. This theoretical and experimental research involves faculty, professional researchers, and graduate students and also draws on the expertise of other departments of the Institute, especially for advanced techniques for physical and chemical analyses. The uniqueness of the investigations organized around the Combustion Research Facility lies in the combination of the human and physical resources of MIT, the coupling of research with advanced education, and the direct involvement of industry in the planning and application of the research.

Figure 1 shows the major combustion-generated pollutants and their effects on acid rain (NO_x, SO_x), atmospheric visibility (volatile organic compounds, NO_x), health (polycyclic aromatic compounds and the inorganic particulates), stratospheric ozone depletion (N_2O), and global warming (CO_2, N_2O). The significant reduction in nitrogen oxide emissions achievable by combustion process modification exemplifies the potential of pollutant emission control through the application of advanced combustion technology. The development of design protocols for reducing NO_x emissions from practical plants is based on improvements in our understanding of the chemistry of nitrogen species and their interactions with hydrocarbons in flames.

Mechanistic pathways of nitric oxide formation and destruction in flames are illustrated in figure 2. The main formation routes include the fixation of atmospheric nitrogen in oxidizing atmospheres at temperatures of 1,600°K and above ("thermal NO") and the oxidation of organically bound nitrogen present in the fuel ("fuel NO"). "Thermal NO" formation can be mitigated by avoiding high peak flame temperatures and minimizing residence times

COMBUSTION GENERATED POLLUTANT EMISSIONS

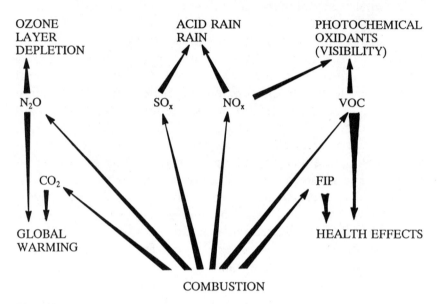

Figure 1

of the combustion products at high temperatures, while the conversion of fuel-bound nitrogen to NO can be reduced by air or fuel staging in the combustion process. Evidence shows that in fuel-rich flames, the fuel-bound nitrogen rapidly forms HCN, which is then converted via intermediates (NCO, HNCO) to other nitrogenous products such as amines, mainly NH and NH_2, with these latter species reducing to molecular nitrogen, N_2. Once the fuel-bound nitrogen has been converted to molecular nitrogen, the rest of the combustion air can be added to burn the fuel completely without a risk of oxidizing fuel-bound nitrogen to NO. The admixing of the combustion air to the burning fuel occurs in stages; hence the name "staged combustion."

A variant of the staged combustion technology involving fuel staging (instead of air staging) permits a reduction of NOx formed earlier in the flame from the combustion of high-nitrogen-bearing fuel such as coal, by an injection of natural gas. The process is known as "NO reburn." The reaction path represented schematically by the right-hand side of figure 2 includes reactions between NO and hydrocarbon fragments such as CH and CH_2. These reactions can convert NO formed earlier in the flame to molecular

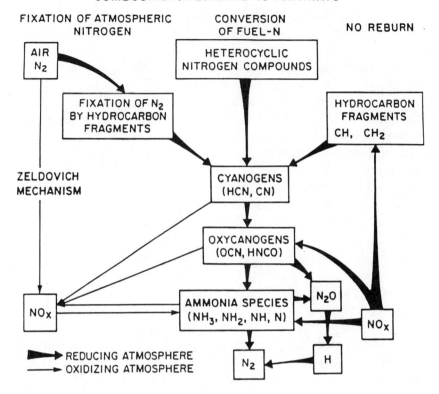

Figure 2

nitrogen and thus reduce the emission of nitrogen oxides by 60–70 percent.

Chemical kinetic modeling based on these mechanistic considerations has helped guide experiments that, in turn, have confirmed the trends shown by computations and permitted the design of low NO_x combustion systems of industrial and utility size.

Unfortunately, the staged combustion of hydrocarbon fuels involving the high-temperature, fuel-rich flame zones that are favorable for reducing nitrogen oxide emissions can also lead to the formation and emission of high-molecular-weight polycyclic aromatic compounds (PACs), which have potentially adverse health effects. Research has uncovered a strong correlation between PAC concentration in the flame and the mutagenic activity of flame samples in bacterial mutation assays. Because PACs must be destroyed with very high efficiency in practical combustion and incineration systems, the air-fuel mixing process in the final oxidative combustion zone must perform to very high standards. In a recent study at the Combustion Research Facility, a method of PAC detection based on laser-induced fluorescence (LIF) has been developed under industrial sponsorship. LIF flame monitoring is done in situ (not requiring a flame sample to be drawn) and has a fast response (on the order of milliseconds); the LIF signal can therefore be used to activate short-time injection of highly oxidizing compounds so as to prevent the emission of trace concentrations of hydrocarbons from incineration processes.

Improving the efficiency of providing power and heat is the most effective possible measure for reducing the emission of all pollutants, in particular that of CO_2. Combined steam–gas turbine cycles have increased energy efficiency because of the favorable matching of the temperature ranges of the working fluids of the respective cycles. Steam cycles operate typically at steam temperatures below 1,100°F, whereas the working fluid temperature in modern industrial gas turbines varies in the range 900–2,300°F. In a standard combined cycle, a gas turbine fired by a premium fuel exhausts into a heat-recovery steam generator that can accept less expensive supplementary fuels such as coal. When a supplementary fuel is used in the heat-recovery boiler, the burden of emissions control falls on the design and operation of the boiler. When supplementary fuels are not used, attention is focused on the gas turbine combustor. A recent development in combustion technology based on the concept of premixing the fuel and air before they enter the combustor to a mixture ratio that corresponds to a gas turbine entry temperature of, say, 2,300°F is presently able to reduce NO_x emission levels as low as 20 parts per million NO_x dry at 15 percent O_2. These ultralean premixed combustion systems hold a promise of even lower emissions from natural-gas-fired gas turbines.

The indirectly coal-fired air heater topping cycle (AHTC), in which coal is used to preheat the air, and natural gas is burned in a topping combustor to raise the temperature further before entry to the gas turbine, provides a prospect of high energy efficiency (greater than 48 percent) with the use of coal and natural gas (typically 60 percent coal, 40 percent gas) at relatively low technical risk. The advantage of this power-generating cycle for reduced CO_2 emissions stems from its high energy efficiency and the partial use of natural gas, a fuel with a lower carbon concentration per unit of energy content than oil or coal. Figure 3 shows the CO_2 emissions

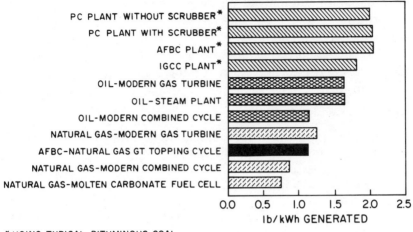

Figure 3

of the AHTC along with those of other power cycles. Emissions of about 2.0 lb/kilowatt hour (kWh) from a coal-fired condensing steam plant drop to about 1.2 lb/kWh for the AHTC. The AHTC offers the possibility of a high-efficiency, clean power-generating cycle fired by coal and natural gas and thus merits an early demonstration project.

Conclusions

Because of the central role of direct fuel utilization in the provision of heat and power for the foreseeable future, major efforts are needed to reduce pollutant emissions from combustion plants. The most cost-effective method for reducing pollutant emissions is the science-based modification of the combustion process. Combustion scientists and engineers at MIT and elsewhere have developed new, clean combustion technologies based on a better understanding of the chemistry of NO_x formation and reduction in flames and the concomitant formation, interconversion, and oxidative destruction of polycyclic aromatic compounds. Improvements in the efficiency of power generation result in the reduction of all combustion-generated pollutants but are of special importance to CO_2 emission reduction because of the very high cost of alternative technologies.

A combined gas turbine–steam cycle in which coal preheats pressurized air and natural gas is used to raise the temperature of the working fluid before entry to the gas turbine merits attention because it promises high cycle efficiency at a relatively low technical risk and because of its partial use of natural gas, which is consistent with both energy conservation and the discriminating use of this premium fuel.

Bibliography

J. M. Beér, The Hoyt C. Hottel Plenary Lecture, 22nd (International) Symposium on Combustion, The Combustion Institute (1988).

J. D. Bittner, J. B. Howard, and H. B. Palmer, in J. Lahaye and G. Prado, eds., *Soot Formation in Combustion Systems and Its Toxic Properties* (New York: Plenum, 1983), 95.

A. G. Braun, P. Pakzaban, M. A. Toqan, and J. M. Beér, *Env. Health Persp.* 72 (1987), 297.

W. F. Farmayan, S.M. thesis, Department of Chemical Engineering, MIT, 1980.

W. F. Farmayan, M. A. Toqan, T.-U. Yu, J. D. Teare, and J. M. Beér, Joint EPA-EPRI Symposium on Stationary Combustion NO_x Control, Boston (1985).

R. W. Foster-Pegg, *Power Engineering* (1982), 80–84..

A. F. Ghoniem, G. Haidervinejad, and A. Krishnan, AiAA/ASME/ASEE 23rd Joint Propulsion Meeting, LaJolla, CA (1987).

M. A. Toqan, J. D. Teare, J. M. Beér, L. J. Radak, and A. Weir, Jr., Joint EPA-EPRI Symposium on Stationary Combustion NO_x Control, New Orleans (1987).

I. M. Torrens, "Global Greenhouse Warming, Joint Role of the Power Generation Sector and Mitigation Strategies," IEA/OECD Seminar, OECD, Paris (1989).

An Environment for Entrepreneurs
Edward B. Roberts

The pursuit of technology-based industrial development has mushroomed in the past decade. Route 128 and Silicon Valley have become prototypes for other regions' and other nations' visions of their own futures. Research and writing about the Technopolis[1-6] have accompanied actions by cities and states throughout the United States and Europe to launch entrepreneurial centers, often based on newly established university incubators and venture capital firms. In Asia, Japan has committed major funding to create a network of "science cities" (going beyond its own Tsukuba), the Republic of China has coupled tax incentives and subsidies to help grow its technology park, and Singapore has linked sophisticated local industrial development planning with government-funded venture capital investments in overseas start-ups to attract high-tech opportunities. Even the Soviet Union has established joint ventures with U.S. and Japanese corporations to generate centers for new technology-based industry.

None of these kinds of governmental programs contributed to the growth of high-technology industry in the Greater Boston area. But what did cause this original American Technopolis to develop? What forces continue today to encourage young local scientists and engineers to follow entrepreneurial paths? In his review of entrepreneurial decision making, Cooper argues for six different potential environmental influences: economic conditions, access to venture capital, examples of entrepreneurial action, opportunities for interim consulting, availability of support personnel and services, and access to customers.[7] This chapter traces the evolution of Boston's high-technology community, providing support for all of Cooper's variables. But I also identify even more critical aspects of culture and attitude that have built a local environment that fosters entrepreneurship.

Early Influences: The Heritage of Wartime Science and Technology

The atomic bomb, inertially guided missiles and submarines, computer-based defense of North America, the race to the moon, and the complex of high-technology companies lining Route 128 outside of Boston are phenomena that became prominent in the postwar years. This was a time marked by a plethora of scientific and technological advances. World War II had defined technology as the critical element upon which the survival of the nation rested. That war brought scientists from the shelter of their labs into the confidence of those in the highest levels of government. And in the postwar years their power and their products and by-products began to shape society, the economy, and the industrial landscape.

How had this started? The sudden need for war research in the early 1940s transformed universities like MIT into elite research and development centers where the best scientific and technological talent was mobilized for the development of specific practical devices for winning the war. Virtually whole universities redirected their efforts from pure scientific inquiry to the solving of critical problems. While many scientists had to neglect their previous research in favor of war-related innovations, the scientists themselves were not neglected. Science and its offspring technology had become the property of the whole nation with an immediate relevance for all the people.

In addition to the urgent expansion and redirection of university research, the war made necessary the reorganization of research groups and the formation of new working coalitions among scientists and engineers—between these technologists and government officials, and between the universities and industry. These changes were especially noteworthy at MIT, which during the war had become the home of major technological efforts. For example, the Radiation Laboratory, source of many major developments in wartime radar, evolved into the postwar Research Laboratory for Electronics. The Servomechanisms Laboratory, which contributed many advances in automatic control systems, started the research and development project that led to the Whirlwind Computer near the end of the war, created numerically controlled milling machines, and provided the intellectual base for undertaking the MIT Lincoln Laboratory in 1951. After the war the Servo Lab first changed its name to the Electronic Systems Lab and continues today as the Laboratory for Information and Decision Systems. Lincoln Lab initially focused on creating a computer-based air defense system (SAGE) to cope with the perceived Soviet threat. To avoid continuing involvement in production and operations once the SAGE system was ready for implementation, MIT spun off a major group from Lincoln Lab to form the nonprofit MITRE Corporation, chartered to aid in the later stages of SAGE and undertake systems analysis for the government. Lincoln then reaffirmed its R&D thrust on computers, communications, radar and related technologies primarily for the U.S. Department of Defense. The Instrumentation Laboratory, growing out of the wartime gunsight work of Dr. Charles Stark Draper—its founder and director throughout his long career at MIT—continued its efforts on the research and development needed to create inertial guidance systems for aircraft, submarines, and missiles. It followed up with significant achievements in the race to the moon with developments of the guidance and stellar navigation systems for the *Apollo* program. The former Instrumentation Lab now bears Draper's name in its "spunoff-from-MIT" nonprofit status. Draper testified as to the scope of these endeavors: "Personal satisfaction ... was greatest when projects included all essential phases ranging from imaginative conception, through theoretical analysis and engineering to documentation for manufacture, supervision of small-lot production, and finally monitoring of applications to operational situations."[8] All these MIT laboratories, major "source organizations" for the new high-technology enterprises that were studied in my entrepreneurship research program,[9] were spawned during a period in which little debate existed about a university's appropriate response to national urgency. These labs have been successful in fulfilling their defined missions, while also providing a base of advanced-technology programs and people for other societal roles.

Building on a Tradition

The war efforts and the immediate postwar involvements of MIT with major national problems, built upon a much older tradition at MIT, enunciated by its founder William Barton Rogers, in 1861, when he created an institution to "respect the dignity of useful work." Its slogan is *mens et manus*, the Latin for mind and hand, and its logo shows the scholar and the craftman in parallel positions. MIT "for a long time ... stood virtually alone as a university that embraced rather than shunned industry."[10] From its start, MIT developed close ties with technology-based industrialists, like Thomas Alva Edison and Alexander Graham Bell, then later with its illustrious alumnus Alfred P. Sloan during his pioneering years at General Motors, and also with the growing petroleum industry. In the 1930s, MIT generated The Technology Plan, to link industry with MIT in what became the first and is still the largest university-industry collaborative, the MIT Industrial Liaison Program.

These precedents were accelerated by the wartime leadership of its distinguished president, Karl Taylor Compton, who brought MIT into intimacy with the war effort just as he himself headed up all national R&D coordination in Washington. In the immediate postwar years Compton pioneered efforts toward commercial use of military developments, among other things helping to create the first institutionalized venture capital fund, American Research and Development (ARD).

ARD was, in part, the brain-child of Compton, then head of MIT. In discussions with Merrill Griswold, Chairman of Massachusetts Investors Trust, and Senator Ralph Flanders of Vermont, then President of the Federal Reserve Bank of Boston, Compton pointed out that some of the A-bomb technology which had been bottled up for four years had important industrial applications. At the same time, it was apparent to Griswold and Flanders that much of New England's wealth was in the hands of insurance companies and trusts with no outlet to creative enterprises. Griswold and Flanders organized ARD in June 1946 to supply new enterprise capital to New England entrepreneurs. [Compton became a board member, MIT became an initial investor, and a scientific advisory board was established that included three MIT department heads.] General Georges Doriot, who was Professor of Industrial Management at Harvard, was later asked to become president.[11]

ARD's first several investments were in MIT developments, and some of the emerging companies, such as Ionics and High Voltage Engineering, were housed initially in MIT facilities, an arrangement that even today would be seen as a source of controversy and potential conflict at most universities. Compton's successor as president of MIT, James R. Killian, furthered the encouragement of entrepreneurial efforts by MIT faculty and staff, as well as close ties with both industry and government. At various times Killian served on the boards of both General Motors and IBM, and as President Eisenhower's science advisor.

The traditions at MIT of involvement with industry had long since made legitimate the active consulting by faculty of about one day per week, and more impressive for its time had approved faculty part-time efforts in forming and building their own companies, a practice still questioned at many universities. Faculty entrepreneurship, carried out over the years with continuing and occasionally heightened reservations about potential conflict of interest, was generally extended to the research staff as well, who were thereby enabled to "moonlight" while being "full-time" employees of MIT labs and departments. The result is that approximately half of all MIT spin-off enterprises, including essentially all faculty-initiated companies and many staff-founded firms, are started on a part-time basis, smoothing the way for many entrepreneurs to "test the waters" of high-tech entrepreneurship before making a full plunge. These companies are obvious candidates for most direct movement of laboratory technology into the broader markets not otherwise served by MIT. Incidentally, few of the faculty founders ever resign their MIT positions, preferring to remain at MIT— as did Amar Bose, founder of Bose Corporation, or Harold Edgerton, co-founder of EG&G—while turning over the full-time reins to their former graduate students and lab colleagues. George Hatsopoulos, founder of Thermo Electron Corporation, Jay Barger, co-founder with another faculty colleague of Dynatech, and Alan Michaels, founder of Amicon, are among the few faculty who left to pursue their entrepreneurial endeavors on a full-time basis, with great success achieved in all three cases.

Although today regional and national governments on a worldwide basis seek to emulate the Boston-area pattern of technological entrepreneurship, in the early years the MIT traditions spread to other institutions very slowly. The principal early disciple was Frederick Terman, who took his Cambridge experiences as an MIT Ph.D. student back to Stanford University, forsaking a faculty offer by MIT to eventually lead Stanford into technological excellence. Terman had gained first-hand exposure to the close ties between MIT and industry, made more important to him by his being mentored by Professor Vannevar Bush, later dean of engineering and then vice president at MIT, who participated in founding the predecessor of the Raytheon Corporation. The attitudes he developed at MIT led Terman to encourage and guide his former students, such as William Hewlett and David Packard and the Varian brothers, to start their high-technology firms and eventually to locate them next to the university in Stanford Research Park.[12] While Terman's efforts obviously produced what has evolved into

"Silicon Valley," where the resulting proliferation of firms came from multiple spin-offs of other companies, and did not follow the dominant Greater Boston pattern of direct fostering of new firms from MIT labs and departments. One early study found only eight out of 243 new technical firms in the Palo Alto area had their origins in Stanford University,[13] probably due in part to Stanford's lack of major government-sponsored laboratories. Indeed, despite the distance from their alma mater, MIT alumni are surprisingly the founders of over 175 companies in northern California, accounting for 21 percent of the manufacturing employment in Silicon Valley.[14]

Our MIT study of major technology-based regions in North America and Europe[15] determined that Research Triangle Park in North Carolina has little evidence of local entrepreneurial activity and few ties between entrepreneurship and the three major universities in that area. And in 1989, only 23 firms in total are documented as "spin-outs" of University of Texas-Austin, including faculty, staff, students, and technology transferred out to other entrepreneurs.[16] Feeser and Willard[17] find far fewer university spin-offs, just one, in their national sample of 108 computer-related founders. Cambridge University, England is seen as heavily responsible for the development of the several hundred high-technology firms in its region, and yet only "17% of new company formation has been by individuals coming straight from the University (or still remaining in it)."[18] Thus, the MIT-Route 128 model remains unusual in its degree of regional entrepreneurial dependence upon one major academic institution. Perhaps other regions need other "models" if they are to achieve technology-based industrial growth.[19]

The Neighboring Infrastructure

Yet MIT has not been alone over the past several decades in nurturing the technology-based community of Boston, now sprawling outward beyond Route 128 to the newer Route 495. Northeastern University, a large urban institution with heavy engineering enrollment and an active cooperative education program, has educated many aspiring engineers who provide both support staff and entrepreneurs to the growing area. Wentworth Institute educates many of the technicians needed to support the development efforts at both the university laboratories as well as the spin-off companies. Boston University and Tufts University, both with strong science and engineering faculties, also play important roles. Even small liberal arts Brandeis University has participated, with Professor Orrie Friedman in 1961 starting Collaborative Research, Inc., forerunner of the much later biotechnology boom in the Greater Boston area.

It may be surprising to hear that Harvard University did not play a substantial role in entrepreneurial endeavors until the recent biotechnology revolution. In many ways Harvard has, over the years, looked down its "classical" nose with disdain at the "crass commercialism" of its technological neighbor a few miles down the Charles River. An Wang, who had worked at the Harvard Computation Laboratory, is the most prominent exception to this rule. Change in regard to encouraging entrepreneurship is in the wind, even at Harvard. The outpouring of excellent research and discovery from Harvard's Chemistry and Biology Departments, as well as from the Harvard Medical School across the river in Boston, has caused Harvard faculty and staff recently to become much more active and successful participants in entrepreneurial start-ups, although not without voiced reluctance and controversy at the University. In fact, in a dramatic revolution of its policies Harvard asked Professor of Biochemistry Mark Ptashne to start Genetics Institute in 1979, a company in which Harvard would hold 15 to 20 percent equity. But protest by critics as to possible influence of such ownership caused Harvard to pull out. Ptashne went ahead and formed the company, while still remaining on the Harvard University faculty.[20] In 1989, the Harvard Medical School took the far reaching step of organizing a venture capital fund to invest in new companies whose founders relate in some manner to Harvard Medical, in some ways mimicking MIT's much earlier activities in regard to ARD, but nevertheless a pioneering step among academic institutions. Indeed, a recent survey of life sciences faculty[21] places Harvard tenth in the nation, with 26 percent, in percentage of "faculty members holding equity in a company whose products or services are based on their own research." MIT life sciences faculty place first in that same survey with 44 percent, for example Professors Alex Rich and Paul Schimmel who co-founded Repligen Corporation. Some of these biotech ventures involve faculty from both Harvard and MIT, such as Biogen, co-founded by Harvard's Walter Gilbert and MIT's Phillip Sharp.

Encouraged no doubt by the exemplary venture capitalist role of Professor Doriot, and separated by a river from main campus influence, many Harvard Business School graduates, joined after its 1951 founding by MIT Sloan School of Management alumni, found welcome homes even in the early company developments. These business school graduates got involved in start-up teams initially as administrators and sales people, and in more recent years participate frequently as primary founders. Thus, Aaron Kleiner, from the MIT School of Management, shares the founding of three high-technology companies with his MIT computer science undergraduate roommate Raymond Kurzweil. And Robert Metcalfe combined MIT educational programs in both engineering and management prior to his launch of 3Com. The Greater Boston environment has become so tuned to entrepreneurship that even student projects with local companies, a part of routine coursework in every local management school, have ended up helping to create numerous entrepreneurial launches. Several firms are claimed to have been generated from feasibility studies done as part of Doriot's famed Manufacturing course at the Harvard Business School. And *Inc.* magazine founder Bernard Goldhirsch credits a Sloan School marketing course with confirming for him the huge market potential for a magazine targeted toward entrepreneurs and small business managers.[22]

Boston entrepreneurs also have benefited from understanding bankers and private investors, each group setting examples to be emulated later in other parts of the country. The First National Bank of Boston (now Bank of Boston) in the 1950s had begun lending money to early stage firms based on receivables from government R&D contracts, a move seen as extremely risky at the time. Arthur Snyder, then vice president of commercial lending of the New England Merchants Bank (now the Bank of New England division of Fleet National Bank), regularly took out full-page ads in the *Boston Globe* showing himself with an aircraft or missile model in his hands, calling upon high-technology enterprises to see him about their financial needs. Snyder even set up a venture capital unit at the bank to make small equity investments in high-tech companies to which he loaned money. Several scions of old Boston Brahmin families became personally involved in venture investments even in the earliest time period. For example, in 1946, William Coolidge helped arrange the financing for Tracerlab, MIT's first nuclear-oriented spin-off company, eventually introducing William

Barbour of Tracerlab to ARD which carried out the needed investment.[23] Coolidge also invested in National Research Corporation (NRC), a company founded by MIT alumnus Richard Morse to exploit advances in low-temperature physics. NRC later created several companies from its labs, retaining partial ownership in each as they spunoff, the most important being Minute Maid orange juice. NRC's former headquarters building, constructed adjacent to MIT on Memorial Drive, now houses the classrooms of the MIT School of Management. Incidentally, long before the construction of Route 128, Memorial Drive in Cambridge used to be called "Multi-Million Dollar Research Row" because of the several early high-technology firms next to MIT, including NRC, Arthur D. Little Inc., and Electronics Corporation of America. The comfortable and growing ties between Boston's worlds of academia and finance helped create bridges to the large Eastern family fortunes—the Rockefellers, Whitneys, and Mellons, among others—who also invested in early Boston start-ups.

And by the end of the 1940s, when space constraints in the inner cities of Boston and Cambridge might have begun to be burdensome for continuing growth of an emerging high-technology industrial base, the state highway department launched the building of Route 128, a circumferential highway (Europeans would call it a "ring road") around Boston through pig farms and small communities. Route 128 made suburban living more readily accessible and land available in large quantities and at low prices. MIT Lincoln Lab's establishment in 1951 in the town of Concord, previously known only as the site of the initial 1776 Lexington-Concord battle of Revolutionary War, "the shot heard round the world," or to some as the home of Thoreau's Walden Pond, helped bring advanced technology to the suburbs. Today Route 128, proudly labeled by Massachusetts first as "America's Technology Highway" and now as "America's Technology Region," reflects the cumulative evidence of 40 years of industrial growth of electronics and computer companies. Development planners in some foreign countries have occasionally been confused by consultants and/or state officials into believing that the once convenient, now traffic-clogged, Route 128 highway system actually caused the technological growth of the Greater Boston area. At best, Route 128 itself has been a moderate facilitator of the development of this high-technology region. More likely the so-called "Route 128 phenomenon" is a result and a beneficiary of the growth caused by the other influences identified earlier.

Accelerating Upward from the Base: Positive Feedback

A critical influence on entrepreneurship in Greater Boston is the effect of "positive feedback" arising from the early role models and successes. Entrepreneurship, especially when successful, begets more entrepreneurship. Schumpeter observed: "The greater the number of people who have already successfully founded new businesses, the less difficult it becomes to act as an entrepreneur. It is a matter of experience that successes in this sphere, as in all others, draw an ever-increasing number of people in their wake."[24] This certainly has to be true at MIT. The earliest faculty founders, Edgerton and his colleagues (the co-founders of EG&G), Bolt and Beranek of the MIT Acoustics Laboratory and then of the company bearing their names (now BBN, Inc.), and John Trump of High Voltage Engineering, were senior faculty of high academic repute when they started their firms. Their initiatives as entrepreneurs were evidences for others at MIT and nearby that technical entrepreneurship was a legitimate activity to be undertaken by strong technologists and leaders. Karl Compton's unique role in founding ARD furthered this image, as did the MIT faculty's efforts in bringing early-stage developments to ARD's attention. Obviously, "if they can do it, then so can I" might well have been a rallying cry for junior faculty and staff, as well as for engineers in local large firms. Our comparative study of Swedish and Massachusetts technological entrepreneurs finds that on average the U.S. entrepreneurs could name about 10 other new companies, three or four of which were in the same general area of high-technology business. Few of the Swedish entrepreneurs could name even one or two others like them.[25] A prospective entrepreneur gains comfort from having visibility of others like himself. This evidence more likely if local entrepreneurship has a critical mass, making the individual's break from conventional employment less threatening.

The growing early developments also encouraged their brave investors, and brought other wealthy individuals forward to participate. As example of the spiraling growth of new firms, even in the early days, Ziegler[26] shows the proliferation of 13 nuclear-related companies "fissioning" within 15 years from Tracerlab's 1946 founding, including Industrial Nucleonics (now Accuray), Tech Ops, and New England Nuclear (now a division of DuPont). With 40 years of activity, a positive feedback loop of new company formation can generate significant outcomes, even if the initial rate of growth is slow. In the mid-1960s, through dramatic proliferation of spin-off companies, Fairchild Semiconductor (founded by MIT alumnus Robert Noyce) gave birth to similar and rapid positive feedback launching of the semiconductor industry in Silicon Valley.[27] And Tracor, Inc. seems to be providing a comparable impetus to new company formation in Austin, Texas, leading to 16 new firms already.[28] Exponential growth starting in the early middle 1970s has generated the several hundred firm Cambridge, England high-tech community.[29]

A side benefit of this growth, also feeding back to help it along, is the development of supporting infrastructure in the region—technical, legal, accounting, banking, real estate, all better understanding how to serve the needs of young technological firms. In Nancy Dorfman's assessment of the economic impact of the Boston-area developments she observes "a network of job shoppers that supply made-to-order circuit boards, precision machinery, metal parts and subassemblies, as well as electronic components, all particularly critical to new start-ups that are developing prototypes and to manufacturers of customized equipment for small markets. In addition, dozens if not hundreds of consulting firms, specializing in hardware and software populate the region to serve new firms and old."[30] Of course, this massive network is itself made up of many of the entrepreneurial firms I have been investigating over the years. Within this infrastructure in the Boston area are new "networking" organizations, like the MIT Enterprise Forum (to be discussed later) and the 128 Venture Group, which serve to bring together on a monthly basis entrepreneurs, investors, and other participants in the entrepreneurial community, contributing further positive loop gain.[31]

This positive feedback phenomenon certainly occurred in the Greater Boston region as a whole and, as illustrated by the Tracerlab example, also was effective at the single organizational level. As one individual or group departs a given laboratory or company to form a new enterprise, the event may mushroom and tend to perpetuate itself among others who learn about the spinoff and also get the idea of leaving. Sometimes one group of potential entrepreneurs feels it is better suited than its predecessors to exploit a particular idea or technology, stimulating the second group to follow quickly. Perhaps as a result, four companies were formed by Instrumentation Laboratory employees to produce "welded module" circuits, a technique developed as part of the Instrumentation Lab's Polaris guidance system project. Ken Olsen, co-founder and builder of Digital Equipment Corporation, recalls that being approached by others to start a company was his first thought about entrepreneurship as a career. The "outside environment" can help this process by becoming more conducive to additional new enterprise formation. In some circumstances, venture capitalists, learning more about a source organization from its earlier spin-offs, may actively seek to encourage further spin-offs from the same source. This certainly played an important role in the 1980s proliferation of biotechnology spin-offs from MIT and Harvard departments.

Other "Pulls" on Potential Entrepreneurs

In addition to the general environmental encouragements on Greater Boston technological entrepreneurship, specific "pulls" are at work on some of the people, making entrepreneurship an attractive goal. Such influences may inhere in the general atmosphere of a particular organization, causing it to be more conducive to the new enterprise spin-off process. For example, until his recent death, Stark Draper, visionary leader of the MIT Instrumentation Laboratory (now renamed the Draper Laboratory), was a key source of encouragement to anyone who came in contact with him. No wonder that the National Academy of Engineering established the Draper Prize to be the equivalent in engineering of the Nobel Prizes in science. With the good fortune to fly coast-to-coast with him one night on a "red eye" from Los Angeles, I learned much about Draper's unique attitudes toward developing young technologists.

I try to assign project managers who are just a bit shy of being ready for the job. That keeps them really hopping when the work gets under way, although the government officials usually want to wring my neck. . . . I break up successful teams, once they've received their honors. That way everyone remembers them for their success, rather than for some later failure. Also, this causes every young person in the Lab to be sitting within 100 feet of someone who's had his hand shaken by the President of the United States. . . . The Lab is a place for young people to learn. Then they can go someplace else to succeed. . . . When I give speeches I single out those who have already left the Lab—to become professors elsewhere, VPs of Engineering in industry, or founders of their own companies. Staying behind in the lab is just for a few old beezers like me who have no place else to go!

His environment was one of high achievement, but with negative incentives for remaining too long. Salaries flattened out quickly, causing the income gap between staying and leaving to grow rapidly as an engineer gained experience. Engineers completing a project had a sharp breakpoint, a good time for someone confident from the success of his or her project to spin-off. In retrospect Stark Draper consciously seemed to encourage spin-offs of all sorts from his laboratory, perhaps the highest attainment achievable by an academic scientist.

They were looking for excitement. They weren't just looking for a more logical way to make software: they wanted to be part of another major breakthrough. After all, Margaret Hamilton had helped send a man to the moon by the time she was 32. "Apollo changed my life," she said. "It had a profound effect on us. Some people never got over it. And there have been other spinoffs from Draper because of it." The follow-up for Hamilton, who was in charge of more than 100 software engineers at Draper, was going to have to be something big. She seems to have found it by starting her own business. To Hamilton, "A growing high-tech company is like a mission." With theory in hand, Hamilton and [Saydean] Zeldin founded HOS [Higher Order Software, the only company in my entrepreneurship research sample founded by two women] in 1976.[32]

No questions were asked if Instrumentation Lab employees wanted to borrow equipment over the weekend, and many of them began their new companies with this kind of undisguised blessing. Draper wanted reasonably high levels of turnover and a constant stream of bright eager young people entering the lab. Over a 15-year period during which I traced "I-Lab" performance, the average age of Instrumentation Laboratory employees remained at 33 years, plus or minus six months. This young-age stability, maintaining the lab's vitality and fighting off technological obsolescence, was not true at most of the other MIT labs studied.

Draper apparently produced similar effects in his teaching activities at MIT. Tom Gerrity, founder of Index Systems, which in turn later created Index Technology and Applied Expert Systems as sponsored spinoffs, reports that Draper's undergraduate elective subject showed him the importance of being able to put together lots of different skills and disciplines to produce a result. Gerrity, now dean of the Wharton School of Business at the University of Pennsylvania, adopted this systems point of view in founding Index several years later, after three MIT degrees and a stint as a faculty member in the MIT Sloan School of Management.

Some other MIT laboratory directors followed similar patterns of entrepreneurial "sponsorship" in smaller, less well-known labs. For example, the head of the Aeroelastic and Structures Laboratory of the MIT Department of Aeronautics and Astronautics had the attitude that the laboratory provided an "internship" type of position and that staff members were more or less expected to move on after a reasonable period. In other labs the environment just seemed to breed entrepreneurism. Douglas Ross, who left the Electronic Systems Lab with George Rodrigues to found SofTech, Inc., comments: "The entrepreneurial culture is absolutely central to MIT. The same mix of interests, drives and activities that make a [Route] 128-type environment is the very life blood of MIT itself. No other place has the same flavor."[33] Ross epitomizes this "life blood" quality. When SofTech was established, MIT took an exceptional step for that time by making a direct equity investment in his ground-zero company, joining a large number of us who shared great confidence in Doug Ross's vision.

Indeed, the challenging projects under way at most of these labs create a psychological "let-down" for their participants when the projects end. Many of the entrepreneurs indicate that they became so involved with their work on a given project that when these projects were completed they felt that their work too was completed. Several of the entrepreneurs attest that their sense of identification with the source lab began to wane as the project neared completion. As with Margaret Hamilton, only through the challenge of starting their own enterprises did they think they could recapture the feeling that they were doing something important.

Beyond the labs, other activities at MIT have over the years encouraged entrepreneurship. The MIT Alumni Association, not the central MIT administration, undertook special efforts to encourage entrepreneurship among its members. Beginning in the late 1960s, the Alumni Association initiated a series of Alumni Entrepreneurship Seminars. Intended to serve an expected small group of 40 to 50 Boston-area young alumni, the effort escalated when over 300 alumni signed up for the first weekend. Over a two-year period the Alumni Association then launched a pattern of weekend seminars targeted for MIT alumni all around the country. Over 2,000 attended the initial national series, and called for more follow-ups. The alumni committee got ambitious and wrote a book on how to start a new enterprise, the only book ever jointly published by the MIT Alumni Association and the MIT Press, and distributed it widely to interested alums.[34] Directories were assembled and widely distributed of alumni interested in the possibility of starting a firm, who might be willing to meet with similarly interested alums, thus beginning a rudimentary matching service. Ongoing monthly programs were started in several cities across the country, including The MIT Venture Club of New York City and then the MIT Enterprise Forum in Cambridge. The latter still continues to stimulate and help new enterprises, and to provide the networking needed to build start-up teams and linkages with prospective investors and advisors. And now the MIT Enterprise Forum has expanded to chapters in 14 major cities across the United States and even in other countries where MIT alumni are concentrated. The Enterprise Forum has recently undertaken operation of the Technology Capital Network, a computer matching service aimed at facilitating linkages between New England entrepreneurs, whether MIT alumni or not, and informal investors (or "angels"). In 1991 the Alumni Association began a new series of MIT Young Alumni Entrepreneurship Seminars, suggesting a renewal of the cycle.

These efforts spread the word, and lend legitimacy to entrepreneurial activities. And they have produced results. Over the years many entrepreneurs have introduced themselves to me, saying they remember hearing me talk years ago at the MIT Alumni Entrepreneurship Seminars. My first meeting with Neil Pappalardo—with whom I much later participated in founding Medical Information Technology (Meditech)—occurred at the first MIT Alumni Entrepreneurship Seminar. Bob Metcalfe, the principal inventor of Ethernet and later the founder of 3Com, a great success in the computer networking market, reports that after attending an MIT alumni luncheon on starting your own business, he resigned from Xerox's Palo Alto Research Center, returned to Boston and established his company with two other engineers.[35] Similarly, the founders of Applicon, now the CAD division of Schlumberger, decided to create their firm after listening to a seminar at Lincoln Lab that reported on the characteristics of the previous Lincoln spinoff entrepreneurs.

And most recently, new policies instituted by John Preston, Director of MIT's Technology Licensing Office, further encourage entrepreneurship, especially by faculty and research staff. In addition to conventional technology licensing to mainly large corporations for fees—which still dominate the MIT technology transfer portfolio—Preston now is willing to license MIT-originated technology in exchange for founder stock in a new enterprise based on that technology. In 1988, the first year of this new practice, six new companies were born based on licensed MIT technology, with 16 firms started in the second year of policy implementation. Matritech is one example, based on technology developed by Professor Sheldon Penman and researcher Edward Fey to employ antibodies to find proteins within cells, a new approach for detecting certain cancers. Entrepreneur Steve Chubb, Matritech's president, received a license from MIT and raised $3.5 million in early outside venture capital in exchange for giving MIT an equity participation in the new venture.[36]

"Pushes" on Entrepreneurship

Some environmental forces affecting the "would-be" entrepreneur are the "negatives" about his or her present employer, rather than the "positives" of going into business. The uncertainties due to the ups and downs of major projects have often been cited as a source of grief, and sometimes even led to expulsion of individuals into a reluctant entrepreneurial path. The evidence suggests that a stable work environment would probably produce far fewer entrepreneurial spin-offs than one marked by some instability. For example, the entrepreneurs who emerged from one large diversified technological firm that I studied rank most frequently "changes in work assignment" as the circumstance that precipitated formation of their companies, followed by "frustration in job." One-fourth of the companies from that firm were founded during the three years that the firm suffered some contract overruns and laid off some technical people, although none of those actually laid off from this firm became entrepreneurs. The "worry about layoff" and seeing the parent firm in a terrible state are cited by many of that period's spin-offs. Even at the Draper Lab, staff was cut by about 15 percent through layoff and attrition after the completion of the *Apollo* program, stimulating a number of new firms. Of the spin-offs from the MIT Electronic Systems Lab (ESL) 92 percent occurred during an eight-year period, when only 28 percent would have been expected if spin-offs occurred randomly over time as a function only of total employment. The large number of ESL projects completed during that period is one explanation for the "lumpiness" of new company creation.

Frustration with the noncommercial environment at the MIT labs and academic departments bothered some of the potential entrepreneurs. Margaret Hamilton, already mentioned in regard to her formation of HOS, exclaims: "The Draper non-profit charter was frustrating, especially if you wanted to get into something exciting. There was always the sense of living in a no-man's land."[37] Many of the entrepreneurs had specific devices or techniques that they wanted to market. Others had no definite products in mind but saw clear prospects for further applications of the technology or skills they had learned at their source organizations. The prospective entrepreneurs usually felt they could not exploit these possibilities at MIT labs, because the labs properly concentrated on developing new technology rather than finding applications for existing technology. Unfortunately for their industrial employers, many of the spinoffs from industrial companies report the same frustration, despite the reasonable presumption that their large firm employers should welcome at least some of these new ideas. In another geographic area, Cooper finds that 56 percent of the new company founders had been frustrated in their previous jobs.[38] Yet frustration should manifest itself more reasonably with just job changing, not company creating, behavior. Clearly the overall environment promoting entrepreneurship in Greater Boston makes the new company option an active choice, if other conditions are right.

And the Beat Goes on

What happens now? At the time of this writing in 1991, the United States and especially the Massachusetts economy are weakened and many observers are in despair about the future of its high-tech industry. Japan, Singapore, Taiwan, and Korea are rapidly growing as centers of technology-based industry. Europe is experiencing increased political and economic consolidation and strengthening. Does all this mean an end to high-technology entrepreneurship in the United States and elsewhere? The evidence suggests the opposite.

At the grass roots, students are showing far more interest in recent years in entrepreneurship courses and clubs. For example, MIT Sloan School of Management students have worked with the MIT Technology Licensing Office to set up projects on possible commercialization of MIT technology. The Sloan School's graduate student New Venture Association has raised money in conjunction with the MIT Entrepreneurs Club, consisting of mainly engineering undergraduates, to provide awards for the best new business plans developed by student teams. At the Harvard Business School, the several elective courses in entrepreneurial management, finance and marketing muster over 25 percent of the total student body, breaking from the HBS traditional concentration on the large corporation. This growth in student interest and enrollment in entrepreneurship subjects is a national phenomenon, manifested in parallel by the outbreak of national academic society meetings devoted to the same topic. Awards for papers or research on entrepreneurship are suddenly being provided by business and engineering schools across the country, further nourishing the interests and exposure. To support these student initiatives, MIT has just approved a proposal to create a Center for Entrepreneurship based in the Sloan School of Management, following the examples of numerous other universities.

Increasingly, states and regions throughout the United States are championing the cause of high-tech entrepreneurship. While Massachusetts was long ago the first state to establish a venture capital organization to aid new firms—the successful and continuing Massachusetts Technology Development Corporation—many other states have joined its ranks, and with much greater political and economic commitment. Pennsylvania's Ben Franklin Partnership has recruited four venture capital firms to begin efforts in different parts of the state, including Zero Stage Capital of Pennsylvania, located in State College, Pennsylvania and working collaboratively with the main campus of Penn State University to transfer technology, build new companies, and stimulate the region's economy. Efforts under way nationwide include targeting state funds or state pension funds to invest in local-oriented venture capital companies, formation of new company incubator organizations in various cities and on college campuses, and developing tax legislation aimed at providing incentives to new or young firms and to their investors. These activities across the United States cross-pollinate and stimulate, not stifle, entrepreneurism in each region.

Indeed high-tech companies are being started at an increasing pace. The Bank of Boston, for example, finds that more new Massachusetts firms were organized by MIT alumni in the past decade than in any prior 10 years.[39] In biotechnology alone they created 20 new companies. A few years back writers like John Kenneth Galbraith came to the wrong conclusion that the age of entrepreneurship in the United States is dead.[40] He argued that only giant corporations could survive in the present era. Today, some writers are equally wrong in arguing that we have too much entrepreneurialism in the United States, which they claim is harmful to our competitiveness in world markets. These modern Luddites urge that government policies should be changed so as to discourage our high rate of new company formation, again claiming that only giant corporations can compete effectively.[41] This naive argument totally ignores that entrepreneurs have been a unique source of U.S. innovation and economic growth for centuries.[42] High-technology entrepreneurs have rapidly moved ideas from university and corporate research and development laboratories out to the market, where both they and society have benefited. And indeed MIT has been the long-recognized international model of this achievement. In contrast, many large corporations have excelled in generating new technologies but have failed to exploit them commercially. The policy issue that does need attention is not how to stifle independent entrepreneurs, but rather how to stimulate comparable corporate entrepreneurship.

One discouraging development for Massachusetts and for the rest of the United States. is the apparent peaking in the late 1980s of new venture capital funds. According to Venture Economics, Inc. new monies added to U.S.-based venture capital (VC) funds declined from slightly more than $4 billion in 1987 to about $2.5 billion in 1989, and increasing fractions of these funds are targeted toward later-stage companies, rather than seed-stage or early growth stage new enterprises. This shrinkage appears to have continued during 1990 and 1991. To put this into perspective, however, as recently as 1980 new commitments to VC funds were only $0.5 billion. The "overshoot" in funding during the six years from 1983 to 1988 in my judgment caused many inappropriate VC investments to be made, and sometimes at rather irresponsibly high evaluations. Money chased deals, and many companies got funded that would have been overlooked in other time periods. (In sharp contrast, Digital Equipment was funded in 1957 with $70,000 that purchased 78 percent of the new firm.) With the exception of these last few years, the new venture capital funds currently available are quite comparable in magnitude with the past. Furthermore, the preference of most venture capitalists has long been for later stage investments, in recent years financing even leveraged buyouts and buybacks of public stock. As a participant in seed funding, I do not notice a dramatically different investment climate at this stage compared to that during most of the last 20 years, albeit the venture capital squeeze might get even worse. From another perspective, the clear institutionalization of the U.S. venture capital industry probably means that it now will go through cycles over time that are comparable to those experienced in the Initial Public Offerings market and in the stock market generally.

Three further aspects of financing deserve comment. (1) Almost all new high-tech enterprises are initially financed by personal savings, family and friends, and informal investors. Later financings still prominently involve the informal investor, alone or in small groups. No evidence suggests that these sources are less available today than in the past. (2) Corporations are today playing a more important role in high-tech financing than in prior years, especially foreign corporations. The increased activity—especially of Japanese firms—in providing early-round funds for U.S. high-technology companies has the favorable short-term impact of more "smart money" being available, with the additional side-benefit to the entrepreneurs of more rapid growth into foreign markets, if not directly at least on a royalty basis. These vigorous direct foreign investments will in the longer run no doubt further strengthen the foreign companies' technological bases, through learning, licensing, alliances, and acquisitions. This foreign-firm strengthening will pose increased downstream competitive problems for larger U.S. corporations that are not actively linking to emerging high-tech enterprises. (3) In terms of other trends, on a worldwide basis more venture capital funds are being developed, financed by the plentiful dollars available abroad, with intended investments partially in their own regions of the world and partially in the United States. In sum, I appraise the overall financing situation for start-up and growth-stage high-tech firms as quite reasonable.

The growth spurt in independent technological entrepreneurship is not limited to the United States. All around the world the symptoms of change are evident. Sticking my neck out a bit I see high-technology entrepreneurship, both in the United States and overseas, as entering a growth mode. From a couple of nodes in the United States, first Route 128 and then Silicon Valley, U.S. hubs of entrepreneurship have spread to Ann Arbor, Boulder, Minneapolis, Austin, Atlanta, Seattle and myriad others. Each area has had its own initiating forces, not all dependent upon a dominant technological university and its laboratories as forbearers. Each has had to go through its own period of start-up, getting to some successes, generating local visibility of role models for others, gradually building financial and industrial infrastructure, and proliferating the positive feedback loops into more active new enterprise formation. This continues to take place throughout the United States, increasingly helped by the role of national media in making the experiences of one part of the country perceived and appreciated by other regions. But high-tech entrepreneurial growth is still primarily a local phenomenon. Only the very beginnings of this pattern are yet under way in Europe and Asia, with a long life ahead. It took over 40 years for what has occurred in Greater Boston to reach its present stage. The next 40 years should see far more technology-based entrepreneurship, both locally and worldwide.

Summary

Although quantitative evidence is lacking to support this assertion, an overwhelming amount of anecdotal data argues that the general environment of the Greater Boston area (beginning during the postwar period), and in particular the atmosphere at MIT, have played a strong role in affecting "would-be" local entrepreneurs. The legitimacy of "useful work" from MIT's founding days was amplified and directed toward entrepreneurial expression by prominent early actions taken by administrative and academic leaders like Karl Taylor Compton and Harold Edgerton. Policies and examples that encouraged faculty and staff involvement with industry and, more important, their "moonlighting" participation in spinning off their ideas and developments into new companies, were critical early foundation stones. MIT's tacit approval of entrepreneurism, to some extent even making it the norm, was in my judgment a dramatic contribution to the Greater Boston culture and economy. Key individual and institutional stimulants like Charles Stark Draper and the MIT Enterprise Forum reinforced the potential entrepreneurial spin-off that derived from a wide variety of advanced technology development projects in MIT labs and in the region's industrial firms. These actions fed into a gradually developing positive loop of productive interactions with the investment community that in time created Route 128 and beyond. Despite near-term pressures upon local entrepreneurialism, the underlying environment seems strong and grass roots activities are growing. Rest assured—while the future is never certain and storm clouds loom for some aspects of technological enterprise, high-technology entrepreneurship remains a continuing and ever more important part of the Massachusetts and the American dream and reality, increasingly shared by aspiring young technologists all over the world.

Acknowledgment

This chapter is edited and reprinted with permission from Edward B. Roberts, *Entrepreneurs in High Technology: Lessons from MIT and Beyond* (New York: Oxford University Press, 1991).

References

1. Nancy S. Dorfman, "Route 128: The Development of a Regional High Technology Economy," *Research Policy*, 12 (1983), 299–316.

2. Roger Miller, "Growing the Next Silicon Valley," *Harvard Business Review*, July–August 1985.

3. Everett M. Rogers and Judith K. Larsen, *Silicon Valley Fever* (New York: Basic Books, 1984).

4. Segal Quince Wickstead, *The Cambridge Phenomenon* (Cambridge, Eng.: Segal Quince Wickstead, November 1985).

5. R. W. Smilor, D. V. Gibson, and G. Kozmetsky, "Creating the Technopolis: High-Technology Development in Austin, Texas," *Journal of Business Venturing*, 4, 1 (1989), 49–67.

6. Sheridan Tatsuno, *The Technopolis Strategy*. (Englewood Cliffs, NJ: Prentice-Hall, 1986).

7. Arnold C. Cooper, "Entrepreneurship and High Technology," in D. L. Sexton and R. W. Smilor, eds., *The Art and Science of Entrepreneurship* (Cambridge, MA: Ballinger Publishing, 1986), 153–167.

8. Charles Stark Draper, "Remarks on the Instrumentation Laboratory of the Massachusetts Institute of Technology," unpublished paper (January 12, 1970), 9.

9. Edward B. Roberts, *Entrepreneurs in High Technology: Lessons from MIT and Beyond* (New York: Oxford University Press, 1991).

10. "A Survey of New England: A Concentration of Talent," *The Economist*, August 8, 1987.

11. Charles A. Ziegler, "Looking Glass Houses: A Study of Fissioning in an Innovative Science-Based Firm," unpublished Ph.D. dissertation, Brandeis University, 1982.

12. Rogers and Larsen, *Silicon Valley Fever*, 31.

13. Arnold C. Cooper, "Spin-offs and Technical Entrepreneurship," *IEEE Transactions on Engineering Management*, EM-18, 1 (1971), pp. 2–6.

14. "MIT Entrepreneurship in Silicon Valley," Chase Manhattan Corporation, April 1990.

15. M. A. Sirbu, R. Treitel, W. Yorsz, and E. B. Roberts, The Formation of a *Technology Oriented Complex: Lessons from North American and European Experience* (Cambridge, MA: MIT Center for Policy Alternatives, CPA 76-8, December 30, 1976).

16. R. W. Smilor, D. V. Gibson, and G. B. Dietrich, "University Spin-out Companies: Technology Start-ups from UT-Austin," in *Proceedings of Vancouver Conference* (Vancouver: College on Innovation Management and Entrepreneurship, The Institute of Management Science, May 1989).

17. Henry R. Feeser and Gary E. Willard, "Incubators and Performance: A Comparison of High- and Low-Growth High-Tech Firms," *Journal of Business Venturing*, 4, 6 (1989), 429–442.

18. Segal Quince Wickstead, *The Cambridge Phenomenon*, 32.

19. Arnold C. Cooper, "The Role of Incubator Organizations in the Founding of Growth-Oriented Firms," *Journal of Business Venturing*, 1(1985), 75–86.

20. "Corporate Album: Genetics Institute," *Boston Business Journal*, March 23, 1987.

21. K. S. Louis, D. Blumenthal, M. E. Gluck, and M. A. Stato, "Entrepreneurs in Academe: An Exploration of Behaviors Among Life Scientists," *Administrative Science Quarterly*, 34 (1989), 110–131.

22. "After the Sale," *Inc.*, August 1990, 39–50.

23. Ziegler, "Looking Glass Houses," 151.

24. Joseph A. Schumpeter, *The Theory of Economic Development* (Cambridge, MA: Harvard University Press, 1936), p. 198.

25. J. M. Utterback, M. Meyer, E. Roberts, and G. Reitberger, "Technology and Industrial Innovation in Sweden: A Study of Technology-Based Firms Formed Between 1965 and 1980," *Research Policy*, 17, 1 (February 1988), 15–26.

26. Ziegler, "Looking Glass Houses."

27. Rogers and Larsen, *Silicon Valley Fever*.

28. Smilor, Gibson, and Kozmetsky, "Creating the Technopolis."

29. Segal Quince Wickstead, *The Cambridge Phenomenon*, 24.

30. Dorfman, "Route 128."

31. Nitin Nohria, "A Quasi-Market in Technology Based Enterprise: The Case of the 128 Venture Group," unpublished paper, Harvard Business School, February 1990.

32. *Boston Business Journal*, August 20–26, 1984, 7.

33. Jane Simon, "Route 128," *New England Business*, July 1, 1985, 20.

34. William D. Putt, ed., *How to Start Your Own Business* (Cambridge, MA: The MIT Press, 1974).

35. Tom Richman, "Who's in Charge Here?," *Inc.*, 11 (June 1989), 37.

36. Udayan Gupta, "How an Ivory Tower Turns Research Into Start-Ups," *The Wall Street Journal*, September 19, 1989.

37. *Boston Business Journal*, August 20–26, 1984, 6.

38. Cooper, "Entrepreneurship and High Technology."

39. "MIT: Growing Businesses for the Future" (Boston: Economics Department, Bank of Boston, 1989).

40. John Kenneth Galbraith, *The New Industrial State*, 4th ed. (Boston: Houghton Mifflin, 1985).

41. Charles H. Ferguson, "From the People Who Brought You Voodoo Economics," *Harvard Business Review*, May-June (1988), 55–62.

42. George Gilder, "The Revitalization of Everything: The Law of the Microcosm," *Harvard Business Review*, March-April (1988), 49–61.

MIT and Industry: The Legacy and the Future
Thomas R. Moebus

Introduction

"...the advancement, development, and practical application of science in connection with arts, agriculture, manufactures, and commerce..."

Such were the goals outlined in the charter granted to the Massachusetts Institute of Technology in 1861 by the Commonwealth of Massachusetts. In accepting this charter as MIT's fifteenth president, Charles M. Vest inherits a legacy of unique cooperation with industry and a challenge to innovate by shaping new partnerships to strengthen MIT's leadership into the next century. With what was in the 1860s an innovative idea—to create a professional school for practicing engineers—MIT embarked on a course that has been marked throughout its 13 decades by a strong and mutually enhancing relationship with industry. MIT's reputation as a knowledgeable yet practical collaborator has often brought industrialists to its doors, while the fertile environment at the Institute has often turned innovators into industrialists, fueling the nation's economic growth. This essay examines the historic roots of MIT's connections with industry, details the present environment of cooperation, and challenges the Institute to a new era of partnership.

In the mid-1870s, word of the pragmatic physics being practiced by MIT Professor Charles Cross attracted a young Canadian to the doors of "Boston Tech." Their collaborative studies of the acoustics of the human ear started the young man on the path to a new speech processing device. His invention patent and later his company's incorporation were filed in Boston. And to this day, MIT's partnership with this man's company remains strong and productive. The young Canadian was Alexander Graham Bell, and in the history of the Bell Telephone Company and later AT&T, over 1,000 MIT graduates have worked to carry out his vision.

In the early 20th century, when MIT was moving from Boston to Cambridge, a then anonymous donor (George Eastman) added a new dimension to the symbiotic relationship between Institute and industry. He specified that MIT would receive only that portion of his pledged donation that it could match with gifts from other firms. Such matching challenges are commonplace today, but at the time they were rare, and this one was meant to carry a specific message: just as industry relied on MIT to produce well-trained engineers from whose labors they derived profit, so too must MIT be able to rely on industry for the resources necessary to carry out its mission. This simple idea of co-reliance is still at the heart of MIT's industrial relationships, though many other symbioses now exist in an increasingly complex field of interaction.

MIT's relations with industry flourished in the 1920s but fell off somewhat during the depression. It was World War II that firmly established MIT as a national resource for knowledge creation and problem solution and set the stage for the next era of industrial interactions. The Vannevar Bush Symposium held at MIT this year reminded us that the boundaries among government, industry, and university almost vanished during the war years, as individual goals coalesced in the pursuit of victory. The flow of people, products, and ideas was almost unhindered. And with a great storehouse of technological achievement available for conversion to civilian purposes, it is no surprise that MIT and industry continued a strong friendship after the war ended.

The Founding of the Industrial Liaison Program (ILP)—1948

In the first few years after the termination of World War II, MIT, like other schools, was faced with a markedly increased need for funds to meet the dual challenge posed by inflation enlarged concepts as to the facilities required for an institute of technology. As now, funds for specific undertakings, often entailing an expansion of the Institute's scope of activities, were relatively easy to obtain, especially from governmental sources. What was especially needed was support given without restrictions as to its use, to enable the Institute to continue the general program to which it was already committed without having to run the risks entailed in deriving too much of its support from the government or any other single source. From this situation evolved a phrase, "funding our independence."

This is how William R. Weems, Director of the Industrial Liaison Program in 1957, described the situation that had given birth to the program a decade earlier. Officially founded in 1948, the ILP was a response to MIT's financial needs. Within a few years it adopted the philosophy of "quid pro quo" that still characterizes it today. The first members of the program came mostly from the oil industry.[1] In the first year of operation, the first ILP symposium (Nuclear Science and Engineering) was held, and 12 industrial visits[2] were hosted by program staff (a part-time director, W. V. Bartz). The ILP grew rapidly. The President's Report listed 26 members in mid-1950, and by 1951 there were 52 companies involved.

The ILP now hosts about a dozen on-campus symposia each year, which draw several thousand visitors to Cambridge, and some 80 off-campus seminar events throughout the world. In addition, over 2,000 visitors from member companies spend a day or two making connections with the MIT faculty. The ILP Publications Office does a monthly mailing to over 13,000 members and others, and distributes some 25,000 MIT papers in response to member requests.

In addition to going "local" with the Associates Program in 1961,[3] the Industrial Liaison Program went international with the enrollment of the Siemens Corporation of Germany in the late 1960s and the NEC Corporation of Japan in the 1970s. The international dimension of its industrial ties has placed MIT, once again, in a leading role in uncharted and sometimes risky waters. I will return to this issue later.

In the early days, very few formal university-industry programs existed at other campuses. But in the 1970s and 1980s the scene changed dramatically. A 1984 study by Kay Tamaribuchi of MIT found over 80 campuses with formal liaison associates and affiliates programs. As budget pressures increased (and as government dollars decreased), such programs were seen as a way to drum up support for higher education from industry. At MIT itself, a number of focused industry exchange programs, called forums or collegia, sprung up to offer liaison-type service in selected technical areas.

Industry-Sponsored Research and Innovations

During the 1960s, federal support for institutions of higher learning was so strong that it temporarily eliminated financial need as a driving force for university-industrial interactions. During this period, there was also an increase of discomfort on campus with the idea of close ties between academia and industry, particularly those segments of industry involved in the military-industrial complex. The share of MIT research sponsored by industry fell in this period to 3 percent.

In the past 15 years, industrially sponsored research has burgeoned at all campuses. During this period, multiclient sponsored research blossomed. Beginning with the Polymer Processing Program in 1978, MIT has been an innovator in this approach. There are now multisponsored programs in all fields of engineering and in management, and there are several programs that bridge the school structure of MIT. A total of 65 programs at MIT now invite industrial sponsorship.

One cannot speak of MIT's role with industry without noting the large number of industrial firms that have been founded within the shadows of the Institute (and at quite remote places) by MIT's faculty, graduates, and students. Several economic studies in the late 1980s identified MIT as the spawning ground for 636 companies in Massachusetts and 176 companies in California.[4] This remarkable achievement speaks to the capacity of MIT as an engine of economic growth. In many cases, such firms and their founders remain close to the Institute, participating as contributors, as governance board members, as sites for student activity, as research sponsors, and in many other ways.

Notable in MIT's history of company formation is a wonderful lack of formality. The formation of companies has been a natural outgrowth of the efforts of MIT's people, not an effect of specific policies formulated with this result in mind. The recent success of the reorganized Technology Licensing Office at MIT, though, shows that institutional efforts can affect the speed with which technologies move into commercialization.

The Present Landscape of Relationships

In the mid-1980s, the Industrial Liaison Program began to craft a new mission in response both to MIT's needs and to our changing view of the needs of industry. ILP redefined its purpose from delivering the most information for the money to "building and maintaining mutually beneficial relationships with corporations." This implies greater efforts at helping corporations maximize the impact of their MIT relationship, in ways defined by the companies themselves. This also implies harnessing the set of MIT-industry relationships for the greater good of the Institute, developing new relationships in priority areas, and assisting faculty to make contacts for intellectual, programmatic, and fiscal support. As the purposes of the ILP were now intertwined with the traditional arena of Corporate Development, the two organizations were united under the banner of Corporate Relations in 1988.

In 1990, the Industrial Liaison Program canvased its 250 member companies in order to increase its understanding of the its clientele. Survey respondents were equally divided among senior executives, directors, managers, and technical staff and fairly represented the program's sectoral and geographic makeup.[5] There was a natural bias toward the research and development end of corporations, from which come the individuals most involved in the program. The survey offered good feedback about the motivation of corporations in building ties with MIT. The primary importance attached to monitoring technology and gaining access to MIT research reflect an ongoing need to extend technical capacities and a desire to take advantage of MIT's broad and excellent technical resources. Building relationships with faculty and recruiting students (most noted by American firms) were also highly rated. The importance given to the professional liaison staff of the ILP offered further evidence that corporations value the person-to-person aspect of such programs. Visits to campus are still seen as a vital means of meeting objectives, with information services also highly rated, but as a support service rather than a main rationale. Surprisingly, continuing education was not seen as a valuable aspect of university relations.

This last finding about continuing education conflicts somewhat with a report on "Industrial Perspectives on Innovation and Interactions with Universities" published recently by the National Academy Press.[6] This survey of 17 industrialists found that "continuing education and cooperative training are essential in supplying them with the capacity to keep their employees educated in the state of the art of technical expertise and equipment." The report reaffirmed the educational role of universities vis-à-vis industrial needs and argued that universities ought not to aim to become major sources of innovation or product development. It emphasized the value of universities and their faculty as "network nodes" for the exchange of ideas and stressed the importance of one-to-one communication among researchers as a key to success.

The ILP members surveyed asserted that MIT was a leader in university-industry programs. Many commented that their firms had withdrawn from other university affiliate programs, a trend also indicated in the "Industrial Perspectives" report. Simple information-exchange programs of an older type may have outlived their usefulness. It is this belief that has motivated the shift in emphasis already noted within the ILP.

Transforming the Partnership between Industry and University

If the market for university-industry programs is in danger of being oversold and underserved, what should be done? A recent study by the Council on Competitiveness urges universities to "develop closer ties to industry so that education and research programs contribute more effectively to the real technology needs of the manufacturing and service sectors." But if such programs are greeted with less interest, or if industries are not willing to supply the resources needed to fund and take advantage of them, what can be done?

What we need are new forms of partnership that recognize the differing aspirations and the differing goals—both short- and long-term—of participants from the two sectors. Given its tradition of leadership in this field and its strong network of contacts in industry, MIT must develop such forms. Global competition, the changing boundaries of companies, and the new communications technologies provide both a challenge and an opportunity to embark on efforts whose purpose is the very transformation of the institutions involved.

Two seminal programs started at MIT in the 1980s offer new frameworks for cooperation with industry. In 1986, MIT President Paul Gray appointed a Commission on Industrial Productivity—a group of senior faculty drawn from departments throughout the Institute—to study and diagnose America's economic performance. What could have been an act of institutional hubris (after all, what did a disparate group of MIT faculty members know about *industrial* productivity?) turned into a brilliant foray across the traditional boundaries of the Institute. The Commission's conclusions were embodied in a very successful book, *Made in America*, and they have been widely praised as a clear-thinking analysis of the American condition. The Commission concluded that major changes are required in the "ways Americans learn, produce, work with one another, compete internationally, and provide for the future," and described strategies for industry, labor, government, and education to accomplish these changes.

Describing strategies is the easy part. Implementing them is the challenge. MIT is now launching a second phase of the Commission study aimed at just that. The exact nature and the tactics of the interaction are yet to be defined, but the goal can be understood as the continuous transformation of both companies and university to keep America effective as a competitive economy. The important first step will be to achieve a high level of joint ingenuity among industry, university, and government leaders applied to a national crisis no less severe, perhaps, than the one we faced in the days of Vannevar Bush.

In a parallel development during the 1980s, and in direct response to the entreaties and complaints of senior corporate executives, MIT developed its Leaders for Manufacturing program. The complaint, in simplified form, was that MIT was not dedicating sufficient intellectual resources to a national problem of growing concern—the competitiveness of its manufacturing industries.

How does an institution with a reputation for turning out technically brilliant individual research performers take on the challenge of educating future leaders for manufacturing industry? The answer has been a vigorous new partnership between industry and MIT that attempts to change the culture of both institutions. Eleven U.S.-based manufacturing firms from different sectors have joined with MIT to sponsor, govern, and participate in the education, research, and leadership training aspects of the Leaders for Manufacturing program. MIT is changing the way it teaches. The companies are changing the way they hire the new leaders MIT is educating. But the biggest change may be in the interface itself.

One might think of Leaders for Manufacturing as a series of "boundary crossings"—each geared to accomplish a goal and to affect the culture on either side of the interface. Within MIT, the Schools of Engineering and of Management collaborate in educating the students in directing and developing the program. Across this technical-managerial interface, the program aims to penetrate the prevailing paradigm of each side with insights drawn from the other. And this culture change is not only about winning "the hearts and minds" of participants, but also about such important dimensions as the tenure process, hiring decisions for new faculty, grading of students, and the like. Teamwork is seen as a key parameter for success.

Within the partner companies, there are attempts to involve not only manufacturing but also design, research and development, and marketing. This is Manufacturing with the "Big M," recognizing that the solution to the problem of manufacturing competitiveness will be found not by focusing solely on how the manufacturing function operates within the company, but rather by integrating a set of concepts about quality, timeliness, and customer responsiveness across the entire corporation.

The boundary between MIT and industry is the one that most concerns us here. Leaders for Manufacturing is in a real sense a joint venture of the partner companies and MIT. Strategic governance occurs through a board composed of senior managers in the companies and senior faculty at MIT. Operational committees, similarly involving both industrial and MIT members, meet monthly to handle tactical issues. Research is conducted both on campus and in company sites by teams of MIT Fellows under joint supervision of faculty and company personnel. Education and curriculum development occur at MIT, but with strong involvement of senior people from the partner companies. One measure of the program's success is the extent to which the partner companies are now sharing the responsibility for second-stage fundraising.

Thus far, Leaders for Manufacturing is on a successful path. Graduates of the program are now finding employment in the partner companies. Research efforts have not only yielded excellent student papers but have in a number of cases led to direct improvements in the manufacturing environment or large savings for the sponsors. This is not to say that the program has succeeded in the larger goal of fully transforming the institutions; but the partners strongly feel that the right steps are being taken and that the model for cooperation is working.

Lifelong Education

What more can be done? While both industrial and university leaders have long recognized the value of "people exchange," comparatively few American companies send their people to participate actively in MIT programs. Culturally, the notion of a staff engineer from anywhere but Massachusetts coming for a long stay at MIT does not cut it. More and more Americans are part of two-career families, and the notion of easily accepted reassignment all over the nation went out with the Organization Man. Individuals who take a year or two "off-line" to collaborate in a research effort at a university may find their job, project, or office gone or occupied when they return. And we all know that the "contract" for lifetime employment, which brings with it a feudal loyalty to the company and a paternal responsibility for human development, is not a specialty of American firms.

In 1982, to commemorate the centenary of MIT's Department of Electrical Engineering (now the Department of Electrical Engineering and Computer Science), a landmark study of "Lifelong Cooperative Education" was conducted.[7] The study reflected a growing realization that "the present rapid rate of scientific and technical innovation invalidates one of the basic assumptions underlying the traditional structure of engineering education: that a few years of formal education can provide an adequate foundation for a lifetime of engineering work." The study noted that the demand for engineers in the United States would increase while the supply would not, and that the rapid rate of technological change would leave an increasing number of midcareer engineers ill-equipped to contribute at the cutting edge of technical competition.

In suggesting that a framework of lifelong education be developed, the faculty study recognized that "any practical solution to the problems...inherently requires the active support and participation of industry." This "necessary intimate collaboration between industry and academe will have to be brought about while preserving the essentially different roles of the two types of institutions" and while recognizing that "the common thread that unites industry and academe is the body of knowledge that is the basis of modern engineering."

The study advances the notion of an "extended campus," a concept intended to blur the distinction between "work" and "study." The faculty envisioned the creation of small academic communities at the workplace paralleled by a great increase in the number of part-time students at engineering schools. A modest amount of role sharing would also occur, so that the distinction between "practicing" engineers and "teaching" engineers would become blurred.

To be honest, we have advanced only a little toward this vision in the past nine years, while the pressures of national competitiveness have increased. But as visions go, one decade may not be too long a gestation period. There are signs that supporting technologies and attitudes may be coalescing in this direction. Notions of the "learning organization" and the need for organizational transformation, for instance, strongly support the kinds of experimentation that are needed.

In discussion with MIT faculty, I have found great sympathy with the simple suggestion that we create part-time but long-term mentorship programs, linking industrial scientists and engineers, faculty, and students. For instance, an industrial mentor might come to campus for two to five days each quarter, with the possibility of longer stays during the summer. While not "overcoming" the cultural hurdles we have noted, this idea could simply sidestep them. For the company, this experience offers an excellent opportunity to build early relationships with students whom they may wish to hire. For students, it offers a more direct view of engineering as practiced in industry. It also offers a low-cost stimulus for midcareer engineers and scientists. The rich environment of MIT is something of a "working vacation" for these individuals, and while many may object to this choice of words, we must take a second look at the long-term implications of such "working vacations" for preventing burnout and obsolescence.

The Virtual Campus

In 1990, MIT's Project Athena concluded an eight-year experiment in campus networking. Athena was a remarkable collaboration in which IBM and DEC each contributed equipment and personnel to assist MIT in creating a workstation environment to support undergraduate education. And while Project Athena has come to an official end, its legacy continues through several offspring: the Center for Educational Computing Initiatives and the Project Athena Consortium. Athena is a learning environment; in its initial form dedicated to the immediate physical campus of MIT. Its true power, though, may lie in its capacity to extend the physical campus via network linkages to participants, independent of their locale—to create or support a location-independent learning environment.

Coordination science—the notion of shared work among groups of geographically separate workers connected by electronic networks—is now a field of study at MIT, under the leadership of Professor Tom Malone in cooperation with a number of firms. The promise of collaborative research and learning within and across organizational borders becomes another element of the new paradigm of electronic education.

Within many corporations, the concept of "the learning organization," as advanced by Professor Peter Senge,[8] is beginning to take hold. Many firms see themselves in the midst of an important transformation, focused on increasing their ability to take in, learn from, and constantly change in reaction to information from all over the world. The value-added by these firms will be in the informational dexterity of their people, implying a vastly increased dedication to learning and human resource development.

Michael Dertouzos, Director of the Laboratory for Computer Science and Director of the Commission on Industrial Productivity, has proposed a major effort to focus attention on the nation's "information marketplace," with a large-scale demonstration project to be led by MIT. In his inaugural address, President Charles M. Vest identified this as one of several important Institute-spanning initiatives for the next decade. Among other results, such an effort would yield a vastly improved system for connecting campuses with each other and with companies, for creating extended learning environments, and for making the vision of the EECS Centennial Committee a "virtual reality."

These efforts and concepts are now unconnected. But they all point to a future of cooperation that will recognize the changing needs and capacities of universities and industry, lead to the creation of new conceptual forms, and technologically support the achievement of the new aims of partnership.

On Internationalism

The blurring of boundaries and the power of information technology to transcend distance have created an international integration of societies and a globalization of business that bring us naturally to examine the goals and implications of MIT's international relations with industry. The postwar years of American prosperity and technological dominance into which the ILP was born have given way to a markedly competitive environment in business and science. In response, MIT has greatly expanded its international contacts and cooperation, its body of foreign graduate students, its research and teaching about international issues, and its concepts of service.

In this context, there has been some criticism of MIT's interactions with firms based outside the United States. To focus a dialogue on this topic, a faculty committee was appointed in 1990 by Provost John Deutch to review the broad range of MIT's international connections—students, visitors, faculty, research, ILP, and foreign support. Professor Eugene Skolnikoff, past Director of MIT's Center for International Studies, chaired this group.

In his inaugural address, President Vest paraphrased the findings of this committee[9] when he proclaimed MIT "a national institution, of and for the United States." The committee's report further defined "MIT's responsibility to the nation" as "served first and foremost by maintenance of its position as a premier institution in education and research in science and technology."

But in recognition of the dual nature of MIT's responsibilities, President Vest strongly committed MIT to an agenda that will continue and expand its international involvement. "We must not endanger the very essence of our institution by retreating into simplistic forms of technonationalism. To draw boundaries around our institution, to close off the free exchange of education and ideas, would be antithetical to the concept of a great university."

It is important to ask just how relationships with foreign firms might change the nature of MIT as an institution. What are the motivations, on both sides, for such linkages? A 1988 faculty study found that a majority of faculty felt a need to keep up with technological advances taking place in foreign laboratories.[10] Those who have visited Japanese industrial laboratories, for instance, found that these visits provided knowledge that had a strong positive impact on their subsequent research. And those who have hosted Japanese visitors in their MIT laboratories felt that the research contributions of these visitors have been significant. This felt need to stay abreast of the world's most advanced research demands that international ties remain and be strengthened.

Further motivation for international ties can be found in the very nature of the problems MIT faculty now address. The global environment, for instance, whether viewed from an ecological or managerial point of view, can only be studied with full understanding of the practices of other nations and of their interdependent impact on each other. Solutions to the most pressing dangers facing us will be found in international cooperation—in science, engineering, and policy—and must involve universities, governments, and industry worldwide.

A third university motivator is financial. A diversified technological institution like MIT requires large infusions of capital to maintain excellence in a broad set of cutting-edge topic areas. As the Institute's reach becomes global, it is natural to assert that its base of financial support should do likewise. Japanese firms have displayed a willingness to support the infrastructure of American universities, including MIT. In fact, there are many outside the ranks of universities who encourage this supportive participation by the Japanese in the world's knowledge creation enterprise.

It is important to keep in mind that the relationships MIT enjoys with firms and other foreign institutions imply a two-way responsibility. While MIT is sometimes criticized for its dissemination of knowledge ("pumping out technology," in the analogy offered by Dean Lester Thurow), it is not given sufficient credit for its accumulation of knowledge ("syphoning in technology," to complete the metaphor). The potential of this capacity has been little exploited to this date; it might, however, include disseminating knowledge about Japanese technology to U.S. industry, or increasing interactions between U.S. and Japanese researchers.

Do foreign companies that develop relationships with MIT and other American universities have special motivations? At the core, we suspect not. But in examining both the ILP Member Survey and the actual practices of firms, some differences emerge. Overall, we note that Japanese companies attach a higher level of importance than do American or European firms to the objectives and activities of university liaison. Experientially, they also invest a great deal of effort in developing the relationship. The traditional channel for developing relationships with U.S. companies has been MIT alumni. Even now, U.S. companies report that recruiting objectives are extremely important as a rationale for supporting university programs. Foreign firms, with few alumni among their ranks, report greater interest in contacts with faculty than with students, and in fact they have not been active recruiters on campus. But patterns of recruiting are slowing beginning to change as more foreign firms establish technical operations in the United States. Similarly, American firms are realizing that they need to develop relationships with universities in other nations. In the long term, though, it seems clear that the lack of a primary university student to industry employee connection will color the nature of MIT's international corporate relations.

Professor Robert Reich of Harvard asks "Who is Us?,"[11] noting the growing incoherence between corporate and national borders. The proliferation of global firms challenges the traditional assumption of unity between corporate goals and national wellbeing.

"What's good for General Motors may still be good for America," but it may also be good for Malaysia; and what's good for Toyota may be good for Maryland. The role of the university in this changing world becomes a bit hazy. Some firms have already asked for MIT's assistance in developing technical and managerial resources all over the globe. Motorola, for instance, is advancing the notion of the "global engineer" whose office could be anywhere and whose field of activity will be wherever there is a project requiring his or her expertise.

One possible adaptation for institutions of higher education is to fulfill the essence of their title as "universities"—that is, to span the "universe" not only in their fields of study but in their student bodies, programs, interactions, and spheres of influence. Few institutions have the resources or capacity to achieve this rank. For MIT, it is a challenge that must be undertaken. MIT's present relationships with industry throughout the developed world, as well as its reputation, offer great leverage for this challenge. In process are efforts to broaden the reach of MIT's business school concepts into new arenas of the Pacific Rim—Singapore, Taiwan, China. The new democracies of Eastern Europe seek both managerial skill and modern technologies to assist them in reaching a sustainable quality of life as market economies. And opportunities for active participation in various regional scientific research programs should be developed and exploited.

A Menu for the Future

Opportunities and challenges abound. While maintaining the independence and separate cultures of university and industry, we must recognize our continued reliance upon each other and the potential of our partnership. With that in mind, I offer some snapshots of a possible new world of MIT-industry interaction:

- The model partnership of Leaders for Manufacturing will be expanded into other arenas, including the environment and the information marketplace. Such programs will be jointly governed by MIT and industry and may also involve other universities. Each program might involve 50 to 100 faculty, over 100 students, and hundreds of industrial collaborators.
- Part-time industry mentorships will abound, with a great increase in particular of American company representatives on campus.

• The MIT Virtual Session, featuring satellite and computer-based courses, will offer a great increase in educational opportunities for practicing engineers and scientists from industry. Such courses may be augmented through residency programs into full professional degree programs.

• The MIT network will include not only the immediate campus community but also the corporate world, facilitating technical inquiries, updates on MIT events, information requests, and direct contacts with faculty.

• MIT's research connections will bridge further into Europe, Japan, and perhaps other nations and will feature inclusion into regional industry-university cooperative programs, providing expanded learning opportunities for students and faculty.

Throughout its history, MIT and its people have been prime innovators in the organizational forms of technology creation and transfer. Willis Whitney, an early product of MIT education and later a professor of chemistry, is credited with inventing the industrial research laboratory at General Electric. Vannevar Bush was one of the early advocates of the private research university and of the notion of involving the university in the knowledge needs of the nation. The MIT Radiation Laboratory, later the Research Laboratory of Electronics, is recognized as one of the world's first interdisciplinary research laboratories. The Instrumentation Laboratory, under the leadership of Charles Stark Draper, was the supreme implementer of Bush's views. And as we have noted, industrial liaison programs were first brought into being at MIT, as were university-based multiclient research programs.

This lead position brings with it a responsibility to develop new forms of partnership with industry and government, to challenge and renew our own institution and the organizations that complete our partnerships, and to find new ways to understand and improve the society and world in which we live and hope to prosper.

Notes

1. Standard Oil of Indiana, Standard Oil of New Jersey, The Texas Company, Socony-Vacuum, Humble Oil, A. O. Smith, Stone and Webster.

2. By comparison, on June 10, 1991, one (out of 19) of our liaison officers hosted visits from six separate firms.

3. The MIT Associates Program offered lower-cost membership to local and smaller U.S. companies. It was merged into the Industrial Liaison Program in 1982.

4. Bank of Boston, "MIT: Growing Businesses for the Future," June 1989; Chase Manhattan Bank, "MIT Entrepreneurship in Silicon Valley," April 1990.

5. Two hundred and forty-seven individuals responded by questionnaire, of which 52 were interviewed by telephone.

6. Industrial Research Institute, *Government-University Industry Research Roundtable* (Washington, DC: National Academy Press, 1991).

7. MIT Department of Electrical Engineering and Computer Science, "Lifelong Cooperative Education," 1982.

8. Peter Senge, "The Fifth Discipline."

9. MIT Faculty Study Group, Eugene Skolnikoff, chair, "The International Relationships of MIT in a Technologically Competitive World."

10. Richard Samuels and Eleanor Westney, "Japanese Scientific and Technical Information at MIT."

11. *Harvard Business Review*, January–February 1990.

The Bear in the Woods
Lester C. Thurow

> There is a bear in the woods. For some people, the bear is easy to see. Others don't see it at all. Some people say the bear is tame. Others say it is vicious and dangerous. Since no one can really be sure who's right, isn't it smart to be as strong as the bear—if there is a bear.
> —Reagan TV advertisement, Fall 1984

Most of the last half century has been devoted to worrying about the Russian bear in the woods. Democracy and capitalism faced off against dictatorship and communism. In the late 1940s and early 1950s it looked as if the Russian bear, helped by the Red Chinese dragon, wished to conquer the world. Aid to Greece and Turkey, NATO, the rearming of Japan and Germany, and the Korean War were all efforts aimed at containing the bears and dragons in the woods.

In the 1950s the Russian bear's military prowess seemed to be matched by its economic and technological capabilities. The Russian Sputnik flew; the American equivalent did not. In the 1950s the Soviet Union was growing faster than the United States. If economic trends were projected forward, the Soviet GNP would pass that of the United States in 1984—a year with ominous literary overtones. In the third world, communism, based on the economic success of the USSR, was widely seen as the only model for economic development. "Containment" was not a problem limited to Eastern Europe. When Khrushchev banged his shoe on the table at the United Nations and threatened to bury the industrial democracies militarily, technologically, and economically, everyone took him seriously. It looked like it was happening.

Kennedy's 1960 campaign for the presidency revolved around "getting the country moving again" on all fronts—militarily, technologically, and economically. With the construction of the Berlin Wall and the Cuban missile crisis shortly after his election, the bear loomed ever larger in the early 1960s. Later in the decade President Johnson spotted an offspring of the red dragon in the woods of Viet Nam. For the rest of the 1960s and first part of the 1970s the dragon's offspring got most of America's attention and resources.

Two oil shocks and the discovery that the Chinese dragon was friendly—if not an ally, at least not an enemy—temporarily diverted attention away from the Russian bear in the mid-1970s. But with a Soviet military buildup in the 1970s (a "buildup" whose reality is now in dispute), the American humiliation in Iran, and the USSR's invasion of Afghanistan, the bear was back—bigger, badder, and more dangerous than ever. In response to the glimpse of this enormous bear in the woods, President Reagan doubled America's military budget in the first half of the 1980s. In addition, he suggested that a huge "Star Wars" program would be necessary to control the bear and its "evil empire."

Suddenly the bear disappeared. The Berlin Wall came down, East and West Germany were united, democracy and capitalism arrived in the formerly communist countries of middle Europe, the Red Army withdrew to the east, the Warsaw Pact was abrogated, and even the Soviet Communist party gave up its monopoly on power. Democracy and capitalism had won. Together they had beaten dictatorship and communism.

What looked like a potential economic superpower in the 1950s had by the 1990s become an economic basket case. Even in military terms, the USSR looks much less formidable in the early 1990s than it did a decade earlier. Ultimately modern military establishments are no stronger than the economies that underpin them.

Economics abhors a vacuum no less than does nature. The economic competition between communism and capitalism is over, but another competition between two different forms of capitalism is already under way. On one side stands the British-American form of capitalism; on the other, the German-Japanese variants. In this competition, all sides claim to be capitalistic, but each organizes its capitalism in a distinctive way.

The king is dead! Long live the king! The game is over! The game has begun!

The sudden onset of a new game creates a psychological problem for the victor of the old game. Victors want to talk about the "end of history,"[1] how they won the old ball game, but there is no time to enjoy victory before one has to get organized to play a very different new game.

When the political thaw was first arriving in the USSR, Georgy Arbotov, head of the Soviet Institute for the Study of the USA, was fond of giving a speech in which he said, "We Russians are going to do the worst thing we can do to you Americans. It is going to be horrible. We are going to deprive you of an enemy! And without an enemy you won't know what to do. You will fall apart!" At this moment the USSR seems more likely to fall apart than the USA, but Mr. Arbotov still has a point. America will not fall apart, but will it recognize that it is in the midst of a very different form of competition and be willing to reorganize to win?[2]

The essential difference between the two forms of capitalism is their stress on communitarian values versus individualistic values as the route to economic success: The "I" of America or the United Kingdom versus "Das Volk" and "Japan Inc."[3]

America and Britain trumpet individualistic values—the brilliant entrepreneur, Nobel prize winners, large wage differentials, individual responsibility for skills, easy firings and quittings, profit maximization, hostile mergers and takeovers. In short, the ideal of the Lone Ranger. In contrast, Germany and Japan trumpet communitarian values—business groups, social responsibility for skills, teamwork, firm loyalty, industry strategies, active government strategies to promote growth. At the level of the firm, the British-American goal of profit maximization is replaced in Japan by what might be called strategic conquest. Americans believe in "consumer economics," while the Japanese believe in "producer economics."

In the British-American variant of capitalism, the individual is supposed to have a personal economic strategy and the business firm is supposed to have an economic strategy that reflects the wishes of its shareholders. They own the firm. For the British-American firm, customer and employee relations are merely a means to the end of higher profits for shareholders. Wages are to be beaten down where possible; and when not needed, employees are to be laid off. Lower wages equal higher profits. On their side, workers are expected to change employers whenever opportunities to earn higher wages appear. They owe their employer nothing. In contrast, many Japanese firms still refer to quitting as "treason."[4]

In communitarian capitalism, individual and firm strategies also exist, but they are built on quite different foundations. The individual's personal strategy is assumed to be more closely bound up with that of the firm in which he or she works. The individual joins a team and is then successful as part of that team. In the

British-American world, company loyalty is somewhat suspect: The individual succeeds as an individual—not as a member of a team.

In both Germany and Japan, individuals are expected to be loyal to the company for which they work. Labor force turnover is seen as bad in communitarian capitalism since no one will plant apple trees (make sacrifices for the good of the company) if they do not expect to be around when the apples are harvested. In contrast, turnover rates are viewed positively in America and Great Britain: Firms are getting rid of unneeded labor when they fire workers, and individuals are moving to higher wage (productivity) opportunities when they quit.

The communitarian business firm must consider a very different arrangement of stakeholders in setting its strategies. In Japanese business firms, employees are the number one stakeholders, customers number two, and shareholders a distant number three. Because employees are the prime stakeholders, higher wages are a central goal of the firm. Profits will be sacrificed to maintain either wages or employment.

Communitarian societies expect companies to invest in the skills of their work force. In the United States and Great Britain, skills are an individual responsibility, and firms are not expected to invest in their employees. Labor is not a member of the team; tt is simply another factor of production to be rented when it is needed and laid off when it is not. Firms exist to promote efficiency by hiring skills at the lowest possible wage rates.

Beyond personal and firm strategies, communitarian capitalism believes in two additional levels of strategizing. Business groups such as the Mitsui Group or the Deutsche Bank Group are expected to have collective strategies. Companies are financially interlocked and work together to strengthen each other's activities. Individual, firm, and business group strategies are then backed up with national strategies. Both Germany and Japan believe that government has a key role to play in promoting economic growth.

In Japan, industry representatives working with the Ministry of International Trade and Industry present "visions" of where the economy should be going. National strategies are orchestrated in key industries. In the past these strategies served as guides to the allocation of scarce foreign exchange or capital flows. Today they are used to guide research and development funding. Microelectronics, biotechnology, materials science and engineering, telecommunications, civilian aviation, robots plus machine tools, and computers plus software are the key targets for the early 21st century. What the Japanese know as "administrative guidance" is a way of life.

Airbus Industries, a civilian aircraft manufacturer owned by the British, French, German, and Spanish governments, is an expression of a "pan-European" strategy. It was designed to break the American monopoly in aircraft manufacturing and get Europe back into civilian aircraft manufacturing. With 20 percent of the world market and a long order book, it is a strategy that has worked. The European landscape is now littered with the acronyms (Jessi, Esprit, Eureka) of government-funded joint R&D projects. Each is designed to help European firms compete in some major industry. Elaborate discussions are now under way as to how pan-European economic "strategies" should be developed and implemented to supplement national strategies.[5]

European governments spend from 5.5 percent (Italy) to 1.75 percent of GNP (Britain) aiding industry.[6] If the United States were to match the German effort (2.5 percent of GNP), it would have to spend more than $140 billion to help its industries in 1991. In Spain, the economy that grew the fastest in Europe in the 1980s, the government owns the firms that produce at least one-half of the GDP.[7] In France and Italy, the state sector accounts for one-third of GNP.[8]

Germany, the dominant European economic power, see itself as having a "social market" economy, not just a "market" economy. German governments (state and federal) own more shares in more industries (airlines, automobiles, steel, chemicals, electric power, transportation—some owned outright, some partially) than any other noncommunist country in the world. Projects such as Airbus Industries have never been controversial political issues. Privatization is not sweeping Germany as it did Great Britain.

Government is believed to have an important role in ensuring that everyone has the skills necessary to participate in the market. Germany's socially financed apprenticeship system is the envy of the world. Co-determination is required to broaden the range of corporate stakeholders beyond traditional capitalists to include workers. Success and failure are not left to spontaneous economic combustion. Social welfare policies are seen as part of a market economy rather than as a regrettable necessity that must be grafted onto a market economy because some individuals fail to provide for their old age or potential health problems. It is believed that markets by themselves generate levels of inequality that are simply too large.

Antitrust and banking laws make group strategies illegal in the American variant of capitalism. In America's economic theology, government has no role in investment funding and a "legitimate" role in R&D funding only in university-based "precompetitive" research, defense, and health care. These rules are sometimes violated in practice, but the theology is clear. In the British-American view, governments should protect private property rights, then get out of the way and let individuals do their thing. Capitalism will then spontaneously combust. Social welfare programs are seen as a regrettable necessity, but there is a lot of discussion about how the higher taxes required to pay for social welfare systems reduce work incentives for taxpayers and how social welfare benefits undercut work incentives for those who receive them.

The Influence of History

These different conceptions of capitalism flow from very different histories. The industrial revolution began in Great Britain. In Britain's formative years during the 19th century, it did not have to play "catch up" with anyone. It was the leader. It was the most powerful country in the world. The United States also had an early and quick start in the industrial revolution. Bordered by two great oceans, the United States did not feel militarily threatened by Britain's early lead. In the last half of the 19th century, when it was moving faster than Great Britain, it could see that it was going to catch up. Extra coal did not need to be thrown into its economic engines to generate more steam.

By way of contrast, 19th-century Germany had to catch up with Great Britain if it was not to be run over in the wars of Europe. "The rulers of German states were expected by their subjects to take an active part in fostering the economic growth of their territories."[9] To have a place at the European table, Prussia had to create a modern industrial economy.

Japan similarly had to develop rapidly if it was not to become someone's (British, French, Dutch, German, American) colony in the last half of the 19th century. In Germany and Japan, economic strategies were important elements in military strategies for remaining independent and becoming powerful. Governments had to push actively to ensure that the economic combustion did indeed take, since it wasn't happening spontaneously. They had to up the intensity of that combustion so that the economic gaps, and hence the military gaps, between themselves and their potential enemies could be erased in the shortest time possible.

At the end of World War II, an intense debate raged as to what should be done about the Japanese and German economies. There were those who argued for the Roman solution—sow the fields of Carthage with salt and permanently destroy its economy. Because Germany was defeated a few months earlier than Japan, systematic deindustrialization was to some extent put into practice there— particularly by the Russians in East Germany.

In the end, what many at the time viewed as an extremely naive "American" approach won the argument. If countries could be made rich, they would be democratic. If their richness depended upon the American market, they would be forced to remain allies of the United States. These naive beliefs underlay the Marshall Plan aimed at the countries (friend and foe) devastated in World War II. It is important to remember that the Marshall Plan was also offered to the Soviet Union and the communist countries of central Europe. The American offer was turned down by Marshal Stalin.

The same ideas led to foreign aid for third-world countries. Prior to the war, the world was dominated by colonial empires, and the purpose of colonies was to make their home countries rich. The purpose of the home country was certainly not to make the colonies rich. While historians argue as to whether colonies contributed to home country wealth or cost more than they were worth, there is no argument as to what the colonial powers were trying to do. The new outlook that resulted in the concept of foreign aid had its successes and its failures, but the successes vastly out numbered the failures. With foreign aid and an open, easily accessed American market, most third-world countries grew from 1950 to 1980 as they had never grown before in all of human history. With the exception of perhaps half a dozen countries, per capita real standards of living were much higher in 1980 than they had been in 1950 in all of the third world.

While the ultimate goal was countries just as wealthy as the United States, probably no one believed that this was really possible in the aftermath of World War II. Naive or not, the systems put in place then—the GATT–Bretton Woods trading system, the Marshall Plan, the European Iron and Steel Community—worked far better than anyone could have believed. Just 45 years later there were several countries as wealthy as America. Some of the third world—Asia's little dragons—were on the verge of making it into the first world. Europe was integrating to create an economic giant. The communist command economies were moving toward capitalism and democracy. The wildest ideas of the naive dreamers of the late 1940s (Truman, Marshall, Monet) were being fulfilled.

In the second half of the 20th century, Germany and Japan were worried about the Russian bear, but they were not charged with containing the bear. Designing, implementing, and financing a strategy to deal with hostile bears or dragons was an American responsibility. Germany and Japan were just loyal allies in the bear and dragon hunts in their neighborhoods. Because of their roles in World War II, no one wanted them hunting in other neighborhoods. Their prime task was not bear hunting; it was to rebuild from the destruction of World War II.

Once a decision was made to encourage economic development in Germany and Japan and they were freed from any responsibility for guarding against marauding bears and dragons, they could devote their energies to economic growth. Building upon the economic muscle of Germany, Western Europe genetically engineered an economic giant in the European Common Market. If the bioengineering can continue with the eventual addition of Eastern Europe and the USSR, the House of Europe could eventually create a body more than twice as large as Japan and the United States combined. In the Pacific, a Japanese economic tiger arose from the ashes of World War II. Emulation led to the birth of four little capitalistic dragons (Korea, Taiwan, Hong Kong, and Singapore) on the Pacific Rim.

As the only undestroyed economic power, America did not need to rebuild after World War II; it had "effortless" economic superiority. As a result, in the present economic competition, Germany and Japan have the advantage that they have long been organizing to strengthen their economic teams. For almost 50 years, they and the rest of the industrial world have been playing a game that Americans have just begun to notice. The United States did not recognize that an economic game between different forms of capitalism was under way because it had military superpower problems to worry about and because it was economically invulnerable. It was not until a large and persistent trade deficit emerged in the 1980s that the country first noticed that a new form of competition had begun.

Economic Warfare?

While an era of intense economic competition lies ahead, it is well to remember that the competitive economic game that the world is about to play is far better than the competitive military game it has been playing. Being aggressively invaded by well-made Japanese or German products from firms that intend to conquer world markets is not at all equivalent to the threat of a military invasion from the Soviet Union or mainland China. Nor does it hark back to the German and Japanese military invasions of World War II. No one gets killed.

Quite the contrary, the competition revolves around such issues as: Who can make the better products? Who can expand its standard of living most rapidly? Who has the better educated and more highly skilled work force? Who can invest more—in plant and equipment, R&D, infrastructure? Who can organize best? Whose institutions—government, education, business—are most efficient? To be forced by one's economic competitors to focus on these factors is good, not bad. And in major ways, the success of any player contributes to everyone's income, since everyone can buy their better products.

Military competition is ultimately wasteful. Resources must be devoted to activities that at best (if they are unused) do not contribute to future human welfare and at worst (if they are used) are destructive to human welfare. Economic competition has exactly opposite characteristics. Everyone is forced to focus on how they can most efficiently make life better for their citizens. "Economic warfare" is not at all equivalent to "military warfare."

From an American perspective, it is also important to remember that being just one of a number of wealthy countries living in a wealthy world is far better than being the only wealthy country in a poor world—even if Americans are sometimes envious of those newly wealthy neighbors, and even if those newly wealthy neighbors sometimes force Americans to rethink how they live.

Before moving on to think about the competition that lies ahead, it is well to spend a moment thinking about the lessons of the competition that is now history. For a brief moment after the fall of the Berlin Wall, it was fashionable to talk about the "end of history." Democratic capitalism had whipped totalitarian socialism, and the world would forever more remain democratic and capitalistic. Were it not for history itself, one might think that such beliefs could only be possible in a country such as the United States, where most live for the present and few look backward.

The belief that the end of history is at hand is, however, fairly common historically. Some who believed it, such as Hitler, did not last very long. Others—the Egyptians, the Greeks, the Romans—all had better reasons (hundreds or thousands of years of continuous history at the top) for believing that their form of social organization would last forever. But they were all wrong. We do not know what it will be, but something will succeed democracy and capitalism as we now know them.

The game has begun!

Notes

1. Francis Fukuyama, "The End of History," *National Interest.* Summer 1989, 4; Richard McKenzie and Dwight Lee, "Should 'The End of History' Have Ever Been in Doubt,?" Center *for the Study of American Business.* working paper 135, October 1990.

2. Michael H. Brest, *The New Competition* (Cambridge: Harvard University Press,1990).

3. Robert Kuttner "Atlas Unburdened: America's Economic Interests in a New World Era," *The American Prospect.* Summer 1990, 90.

4. "Graduates Take Rites of Passage into Japanese Corporate Life,"*Financial Times*, April 8, 1991, 4.

5. Hans Dieter Weger, ed., *Strategies and Options for the Future of Europe: Aims and Contours of a Project.* Bertelsmann Foundation, 1989.

6. Robert Ford and Wim Suyker, "Industrial Subsidies in the OECD Countries," *OECD Economic Studies*, No. 15, Autumn 1990, 37.

7. "State Still Accounts for half of GDP," *Financial Times*, March 15, 1991, Spain 4.

8. "Sell by 1992," *The Economist*, March 20, 1991, 14.

9. W. O. Henderson, *The Rise of German Industrial Power, 1834–1914* (Berkeley: University of California Press, 1955), 71.

Appendix

Inaugural Address, May 10, 1991
Charles M. Vest

This is, indeed, a splendid moment—as we gather to celebrate a great institution, to renew our commitment to a set of ideals, to mark a passage, and to set a course for the future. Yet for me, and for my family, it is also an intensely personal experience, and one that we are honored to share. A journey that began in a warm family in a small town in West Virginia has led to center stage in Killian Court—where my own path, and that of the Institute, have come together in this symbolic moment. It is a profound privilege to walk with four great and gracious men—Jay Stratton, Howard Johnson, Jerry Wiesner, and Paul Gray. Your trust and guidance give me great comfort and courage for the task ahead.

On the banks of the Charles River an institution has arisen that is recognized throughout the world for its unique contributions to our life and times. Established 130 years ago this spring, MIT did not become yet another comprehensive university. Nor did it become simply an "engineering school" or a "polytechnic institute."

Rather, it became a wellspring of scientific and technological knowledge and practice, and a place where musical creativity thrives. Its inventive and entrepreneurial faculty generated a great economic engine, and they have created revolutionary insights into the structure of language and the nature of learning. They have led the quest to decipher the molecular foundations of life, and they have influenced the political and economic policies of nations. MIT's engineers and scientists made critical contributions to our nation's security when that was largely a military matter, and its graduates have given architectural manifestation to humankind's highest cultural and artistic insights.

MIT has been home to distinguished scholars from around the world, men and women who have stretched the human mind and spirit. Above all, it has provided an intense and effective education to generations of the brightest young men and women that this nation, and the world, have brought forth.

Now MIT prepares for the passing of the 20th century and the advent of the 21st. We seek form and substance appropriate for these times, even as we seek to shape the future of our nation and world.

But we enter more than a new temporal era. We stand at the dawn of a new global age. Our lives are interwoven across national boundaries in unprecedented ways—connected through our earth's environment, whose stewardship we all share, through our economic and production systems, through instantly shared information, through universally shared dreams. These dreams include the vision of a world in which the security of nations is defined by economic and social dexterity rather than by military might. And they include the vision of a nation that has regained its sense of social justice and is truly the land of opportunity for all.

MIT has played a remarkable role, at critical moments, in shaping our nation and our world. We have done so through individual creative genius and through grand institutional ventures. Like America itself, we have responded in a heroic and innovative manner to sudden challenges, such as the onset of World War II or the launching of Sputnik. Today we are challenged once again on a grand scale. But this time by slow, corrosive forces rather than by sudden, galvanizing events. By the erosion of our global environment rather than by explosions at Pearl Harbor. By declines in scientific literacy and industrial competitiveness rather than by the launching of a satellite.

This morning I would like to share with you my view of the challenges that confront us and to offer a growing vision of the opportunities they present for the future of MIT.

A New Global Age

There is a remarkable image etched in the mind and psyche of our generation. We were the first to view a shimmering, seemingly peaceful planet Earth from the depths of space. Still, here below, we know that we inhabit a raucous global village. We are connected, across time and space, as never before in human history. Many of these connections have been made possible by the advances in science and technology. We must learn to deal with this interdependence in new ways, creating new forms of organization and incorporating new points of view.

Let me give three examples.

First, the earth's environment: a fragile envelope that bears witness to the degrading effects of human activity. It is no longer possible, if it ever was, for individuals or nations to think that the way in which they treat their land, air, and water has no bearing on their neighbors. Nor is it possible for us to work on each aspect of this damaged environment as a separate problem. Ironically, many of the scientific and technological advances that so enhance human comfort and well-being—advances in transportation, energy, and agriculture—concurrently pose threats to our biosphere. This presents a challenge and an opportunity for us here at MIT. I believe that we must marshal our interests and capabilities to understand these issues and to develop solutions. Such an endeavor will require a new generation of scientific computation for atmospheric modeling, new instrumentation for monitoring environmental conditions, new modes of analysis, and new technologies to correct or avoid problems.

Beyond this, we need to come together in new ways—from different fields, different organizations, and different countries—to understand not only the physical, but the cultural, economic, and political forces that affect the health of the natural world. The stage has been set at MIT by the establishment of the Center for Global Change Science and by the new Council on the Global Environment. Only with this kind of integrated approach—drawing on faculty from disparate fields—can we hope to meet the profound challenge of making and keeping our planet livable.

Another challenge—and set of opportunities—in our increasingly interdependent world lies in the realm of electronic communication. Instantaneous communication, both verbal and visual, has reduced our planet to the electronic global village once envisioned by McLuhan. Knowledge has become a capital asset, at least as important as physical resources. Bits of information flowing through copper wires, optical fibers, or satellite links have become a new currency: the currency of the information marketplace. Increasingly, the commerce of this new marketplace will be conducted along fiberoptic information superhighways that will connect computers, telephones, high-definition video systems, and hybrid technologies yet to be developed.

This information infrastructure already exists in rudimentary form. MIT has an opportunity to play a pivotal role in bringing increased capabilities and coherence to this system and in defining the currency of the new information marketplace. In doing so, we must not only increase the power and ease of computing and communications, but we must do so in ways that enhance our intellectual and social capabilities, that help us make wiser decisions, and that enable us to bridge cultural and political barriers. Here, too, we must invent new ways of combining our talents across disciplinary and institutional boundaries in order to give form, substance, and humanity to the dawning information age. To this end, I am pleased to announce the establishment of the MIT Information Infrastructure Initiative—a project that will bring together eight different organizations within MIT with the goal of working with industrial partners to develop a very high frequency, entirely optical network and to establish within our campus a working model of the information marketplace.

My third example derives from the increasing political and economic connections throughout the world. These connections pose the question of whether the MIT of the future will be a national or an international institution. What does it mean for MIT to be a citizen of a world where common problems or interests are often more powerful than geographic distances, yet where national differences exist?

The issue is complex. MIT is a national institution. But America is no longer isolated. MIT was born as a manifestation of Yankee ingenuity and know-how, it has served as a driving force for the creation and improvement of American industry, it is funded to a very significant extent by the American taxpayer, and above all it is centered on the education of many of the brightest and most talented young people of the United States. MIT is of and for America.

Today, however, in order to serve America well, we must participate in the broader global community. Basic science has always prided itself in being the prototype for true international cooperation, but today this viewpoint and system are being strained—strained because of the increasing economic value of university-generated knowledge and technological concepts.

There are those who look at this country's position on the economic balance scales and call for greater protection of our ideas, especially those having to do with science and technology. Some look at this country's troubles in the world marketplace and are quick to blame our overseas competitors. Others cast the issue into the framework of Pogo Possum's famous saying: "We have met the enemy, and he is us." And still others quickly respond along the lines of Robert Reich, who asks, "Who is us?"—that is, in this day and age, what defines an American corporation?

Clearly, we must be concerned with this nation's economic well-being. We must not, however, endanger the very essence of our institution by retreating into simplistic forms of technonationalism. To draw boundaries around our institution, to close off the free exchange of education and ideas, would be antithetical to the concept of a great university. The list of nations that, at difficult historical moments, closed their universities to the outside world is not one we would be proud to join.

This does not mean that we could not, on occasion, establish special programs directed at the solution of national problems. However, any such programs must also fit one fundamental rule: All students, once admitted to MIT, must be able to participate fully in our educational and research programs, without regard to their citizenship.

In my view, a much more important concern of MIT should be the establishment of programs to ensure that our students are educated in such a way as to prepare them to lead full, responsible lives as world citizens. It is time we made the matter of international context and opportunity an integral part of an MIT education.

The Changing Face of America

Just as we develop new connections among nations, so too must we seek new connections within our own. The face of America is changing significantly and rapidly. Our society is increasingly pluralistic, yet our connections across racial, ethnic, and sometimes even gender boundaries are frayed. Securing America's promise for all remains a crucial goal. The nation's potential will not be realized until all racial and ethnic groups have full opportunity to realize their own potential and, in doing so, to contribute to the health and vigor of our society.

MIT has traditionally educated engineers, scientists, and others to develop technologies, lead businesses, and serve as professors, researchers, and scholars. To continue this leadership in the era ahead, we must better reflect the changing face of America in our students, faculty, and staff.

We can clearly see such changes in our undergraduate population—thanks to the leadership, commitment, and concerted effort of many here with us today. Among our graduate students and our faculty, however, we see far less evidence of this change as yet. We must double and redouble our efforts to attract the brightest and best from all races, both women and men, not only to our undergraduate program, but to our graduate school and to our faculty. There are many social and historical forces working against success in this endeavor. It will require renewed commitment on the part of each of us to identify and recruit these scholars and, once they are here, to do our part to see that they attain their full potential.

As one step, we will begin implementing during the coming weeks a program proposed by the Equal Opportunity Committee to recruit more women to our faculty. And we will reaffirm and reinvigorate our policies and programs for bringing more underrepresented minority members to our faculty. As we succeed, and in order to succeed, with these and other efforts, we must work to ensure that MIT is a place that respects and celebrates the diversity of our community. Just as we celebrate learning about the physical universe, or the political and economic worlds or the creative arts, so must we celebrate learning about, and from, each other. Such change is rewarding, but it is seldom easy. During the years ahead we must refuse to let the centrifugal forces of intolerance and injustice pull us apart. We must be held together by respect for the individual and by a commitment to the values we hold in common.

Education: To Move a Nation

Just as we as individuals are part of an interwoven social fabric, so too is MIT part of an interdependent educational system—one that begins before kindergarten and extends through postdoctoral studies. Within this system, America's colleges and universities stand as national treasures. But the strength of these institutions, and thus of our society, is imperiled—imperiled by the state of our primary and secondary schools, and imperiled by the declining interest and ability of our young people in the pursuit of rigorous advanced studies, particularly in science and engineering. These trends must be reversed.

It is my firm belief that national educational strength is the essential prerequisite for economic and social prosperity. Education can move a nation: The future belongs to those who understand it. At all levels, active, informed participation in our economy and our democracy now requires an ability to understand basic scientific and technical concepts. And yet American popular culture pushes us in the opposite direction. We need no less than a change in the culture of this country, a revolution in attitude about the importance of education and, in particular, of scientific and mathematical literacy.

Until we, as a nation, wake up to the fact that we must increase our investment in the growth of human capital—that is, people and ideas—our educational system will spiral downward, pulling our economy and our way of life with it. This is a danger of the first magnitude, and we must all work to address it.

Thirty years ago, MIT played a key role in launching a nationwide wave of education reform in the sciences. The time has come again for us to place our expertise and stature in the service of a major national effort to rebuild the strength of science and mathematics in American schools. I believe that MIT not only can, but must, draw on its special strengths to help renew effective, accessible education for the young people of this country.

An MIT Education for the Future

The education that we most directly influence, however, is the education of our own students. Among them are people whose passion is to engineer a better world. Among them are people with a particular, concentrated brilliance. Among them are profoundly creative people who tread new and different pathways. We are gifted with some of the very brightest young people of our nation and of the world. It is through these students that MIT will have its greatest influence on the world of the future.

In recent years, our faculty has been involved in a long-term review of the undergraduate program. The intensity of this review is testimony to the fact that education, and particularly undergraduate education, is at the very core of MIT. No one has been more engaged with these matters over the years than our engineering faculty. Indeed, the engineering curriculum in this country was largely developed by MIT faculty in the 1950s and 1960s. They spearheaded the infusion of basic science into engineering education and practice.

The results were astounding: We produced engineers who created a revolution in computing and communication, developed vehicles to explore outer space, and started not only companies, but entire industries based on high technology. While this curriculum has been continually refreshed, its fundamental approach and content have remained essentially unchanged for thirty years. The world in which engineering is practiced, on the other hand, has changed dramatically and rapidly.

Take, for example, the decline in the ability of the United States to compete in the world marketplace for manufactured goods. The reasons for this decline are complex, but a major issue has certainly been the attitude of industry and of universities toward the design and manufacture of consumer products. We need to infuse our engineering students with an increased respect for and enjoyment of effective, efficient, and socially responsive design and production. Today, we must prepare engineers who have the self-discipline, analytical skills, and problem-solving abilities that are so highly valued in MIT graduates, but who are also prepared for the challenge of production and leadership in the world marketplace of the next century.

This is but one of the challenges to engineering education. But it is indicative of the concerns that face our faculty as they design a curriculum that will serve our students well into the 21st century. They will do so in the setting of this research university: a setting in which the unique blending of graduate education, undergraduate education and research creates unparalleled opportunities for learning and for discovery—a setting that keeps both our education and our research forward-looking and robust.

All do not agree with this view. Many believe that our mission has become distorted and that education has been lost in our desire and responsibility to excel in research. This is clearly a central issue for MIT—one that must be openly discussed in all corners of the Institute. This fall, as an event of the inaugural year, we will hold a major colloquium on the topic of teaching and learning within the research university. I intend this to be a no-holds-barred debate that will illuminate our efforts to shape the future of education at MIT.

Educational success at MIT depends, above all else, on the commitment and inventiveness of our faculty. Excellence in undergraduate teaching must be rewarded and encouraged. To this end, we are establishing an endowed program to recognize faculty members who have profoundly influenced our students through their sustained and significant contributions to teaching and curriculum development. A select number of faculty will be appointed as Faculty Fellows, each for a ten-year period, and will receive an annual scholar's allowance throughout their appointment. The first Fellows will be appointed this year, and we expect their ranks to build to at least sixty during this decade.

The strength of an MIT education is its depth and intensity. Our graduates value above all else their self-discipline, analytical thinking skills, and confidence to take on great challenges. Today, science and technology, culture and policy, industry and government, production and communication, are interwoven as never before. The nation needs broadly educated young men and women to be leaders of the next generation. An understanding of science and technology is surely part of what such leaders must possess. Similarly, those who practice science and technology need an ever greater understanding of the world in which they will work and must be able to contribute wisely to policies affecting the development and uses of technology.

What does this mean for education at MIT? Surely it means a careful balance among the humanities, arts, and social sciences on the one hand and mathematics and the physical and life sciences on the other. And it means a continuing look at our departmental programs to ensure that—in content and approach—they give our students the best possible foundation for intellectual growth and professional achievement.

Our campus should be a place where humanistic and artistic scholarship and creation can flower in unique and important new ways. I further believe that we at MIT have an unusual opportunity to create an environment in which the humanities and engineering enrich each other. While the continuum from the humanities to the natural sciences has long been recognized, the continuum from the humanities to engineering is less well explored. In general, such exploration has been hindered by a utilitarian view of the humanities and social sciences on the part of many engineering educators, and by a lack of appreciation of the intellectual content of modern engineering by many humanists. An MIT education should enlarge an individual's choices—and so should include a common experience in science and mathematics, a serious exploration of the humanities, arts, and social sciences, and a continuing conversation among these fields.

I believe that the creative tension generated by these varying interests and cultures can serve us well as we continue to review and renew our undergraduate programs. We have a common currency of excellence and creativity—regardless of field—that will enable us to develop new modes of inquiry and teaching that make the most of the unique intellectual community that is MIT. We have a special set of talents and focus that give MIT its distinctive character. By building on its special strengths, MIT will contribute in rich and often unique ways to the times and the nation's needs.

We should not expect to be all things to all people. One of the great strengths of the American educational system is the great variety of its public and private colleges and universities. This condition allows for, and indeed demands, experimentation, variation, cooperation, and competition. The resulting synergy is the yeast that keeps our system strong.

Rebuilding Trust in Science and Technology

For four decades, the American research universities have served this nation exceedingly well. From virtually any perspective, they have paid enormous dividends in return for the public's trust and investment. Dividends in the form of educated leaders in academia, business, and government, advances in medical care and nutrition, national security, new and revitalized industries, and increased understanding of our physical, social, and natural worlds. But today the American public is calling into question the value of our research universities and no longer tends to view science and technology as the foundation of progress. The public's attention is caught not only by the debate over the costs and quality of undergraduate education but by the debate over the costs and conduct of research.

The doubt of the moment, however, must not be allowed to weaken the basic concept of the American university system, one that is universally recognized as the best in the world. This system is founded on a social contract with the American public and enhanced by partnerships with government and industry. We cannot keep our flexibility, our vigor, our quality—as a nation or as an academic community—by taking this partnership for granted. We need to rebuild trust in this nation's research universities and its scientific enterprise. We must ensure that the foundation of scientific and scholarly research is secure. What is this foundation? Jacob Bronowski stated it with deceptive simplicity when he wrote, "The end of science is to discover what is true about the world."

In seeking scientific truth, ideas and hypotheses are debated, tested, proved, disproved, revised, built upon, or rejected. This activity is carried out by researchers in different laboratories, in different universities, indeed in different countries. This is what makes science, indeed most scholarship, simultaneously an individual and a communal activity. And it is why we have usually been able to rely on this system to detect and correct error. Like all human endeavors, science is not, and cannot, be totally free from error or even occasional abuse. And so it rests upon us—as scientists and scholars—to do a better job of strengthening, continually renewing, and transmitting our system of values.

Great teachers impart and stimulate the passion, excitement, and beauty of intellectual endeavor. But it is equally important that we impart and stimulate the meaning of, the necessity of, and the passion for the pursuit of truth with integrity and ethical rigor. But whatever we say, ethical lessons will be taught primarily by the ways in which we undertake our own scholarly activities.

These lessons will also by conveyed by the ways in which our institutions handle problems if they do arise. How we deal with alleged misconduct will also affect the strength of society's confidence in and regard for our universities and colleges, and for the enterprise of science. We have heard great outcries—for and against—the policing of science. Our response, as an academic community, must not be one of knee-jerk defensiveness against our critics. Rather, we must engage seriously with our thoughtful critics as well as with our colleagues as we develop ways to continuously foster academic integrity and deal forthrightly and fairly with problems when they arise. If we are not able to do so, we can be sure that others will be only too glad to do it for us.

Public confidence in our universities must be fully restored. Our social compact must be reestablished. But in the discourse required to do so, we must avoid the trap of justifying all that we do on utilitarian grounds. Clearly, we have been great contributors to the nation's economy, and this must continue to be a cardinal element of MIT's mission. But we must take care not to overemphasize these contributions as the justification for investing in universities. If we overuse such arguments, we might unwittingly endanger our traditions of intellectual excellence, innovation, integrity, openness, worldwide service, deep scholarship, and independent criticism. Ultimately, our contributions to social progress and well-being rest on our ability to steer our own course, with imagination and intellectual daring.

MIT: Shaping the Future

What, then, is my vision of MIT a decade hence?

MIT will be a preeminent wellspring of scientific knowledge and technological innovation. MIT will foster the pursuits of individual scholars, whose work so often leads to truly fundamental discoveries. We will be known for our ability to establish new and effective methods for analyzing complex and pervasive problems facing the nation and the world. In an invigorated partnership with industry, the government, and other educational institutions, we will contribute profoundly to their solution. MIT will be known for educating engineers who combine the spirit of innovation and invention with a passion for the highest quality and efficiency in design and production.

MIT will better reflect in our students, faculty, and staff the changing face of America. We will find ways to instill the excitement and romance of science and mathematics in new generations of young people. MIT will spearhead efforts to rekindle our nation's belief in the importance of scientific research and education. We will have found renewed commitment to the deepest values of the academy. MIT will stand for integrity in all that it does. MIT will serve our nation well, but also will be of and for the greater world community.

Above all, the Massachusetts Institute of Technology will be a place to which the brightest young men and women will come for their educations. They will be able to attend MIT regardless of their financial circumstances. They will be taught and counselled by dedicated teachers who themselves define the leading edge of human knowledge and invention. Their education will be robust: deep in scientific content, yet providing the flexibility and learning skills to serve them well in ever changing and expanding circumstances. They will be attuned to the complexities of their world, a world that they will help to change. Through that wonderful blend of undergraduate education, graduate education, research, and creative activity that is MIT, our students will be enriched, and they, in turn, will enrich the Institute.

Mens et manus: With mind and hand we set forth. Our promise will be secured by the collective energies and wisdom of those who are drawn to this great magnet for intellect and creativity. Together, we will give shape to the future—the future of MIT, our nation, and our world.